精通Scrapy
网络爬虫

刘硕 编著

清华大学出版社
北京

内 容 简 介

本书深入系统地介绍了 Python 流行框架 Scrapy 的相关技术及使用技巧。全书共 14 章，从逻辑上可分为基础篇和高级篇两部分，基础篇重点介绍 Scrapy 的核心元素，如 spider、selector、item、link 等；高级篇讲解爬虫的高级话题，如登录认证、文件下载、执行 JavaScript、动态网页爬取、使用 HTTP 代理、分布式爬虫的编写等，并配合项目案例讲解，包括供练习使用的网站，以及京东、知乎、豆瓣、360 爬虫案例等。

本书案例丰富，注重实践，代码注释详尽，适合有一定 Python 语言基础，想学习编写复杂网络爬虫的读者使用。

本书封面贴有清华大学出版社防伪标签，无标签者不得销售。
版权所有，侵权必究。侵权举报电话：010-62782989 13701121933

图书在版编目（CIP）数据

精通 Scrapy 网络爬虫/刘硕编著. —北京：清华大学出版社，2017（2018.6重印）
ISBN 978-7-302-48493-6

Ⅰ. ①精… Ⅱ. ①刘… Ⅲ. ①软件工具－程序设计 Ⅳ. ①TP311.561

中国版本图书馆 CIP 数据核字（2017）第 230263 号

责任编辑： 王金柱
封面设计： 王　翔
责任校对： 闫秀华
责任印制： 李红英

出版发行： 清华大学出版社
　　网　　址： http://www.tup.com.cn, http://www.wqbook.com
　　地　　址： 北京清华大学学研大厦 A 座　　**邮　　编：** 100084
　　社 总 机： 010-62770175　　　　　　　　　　**邮　　购：** 010-62786544
　　投稿与读者服务： 010-62776969, c-service@tup.tsinghua.edu.cn
　　质 量 反 馈： 010 62772015, zhiliang@tup.tsinghua.edu.cn

印 装 者： 清华大学印刷厂
经　　销： 全国新华书店
开　　本： 180mm×230mm　　**印　张：** 14.5　　**字　数：** 325 千字
版　　次： 2017 年 10 月第 1 版　　　　　　　　**印　次：** 2018 年 6 月第 3 次印刷
印　　数： 5001~7000
定　　价： 59.00 元

产品编号：074439-01

前　　言

关于本书

如今是互联网的时代，而且正在迈入智能时代。人们早已意识到互联网中的数据是有待开采的巨大金矿，这些数据将会改善我们的生活，网络爬虫开发工作岗位的出现和不断增加正是基于对数据价值的重视。优秀的爬虫框架就像是开采金矿的强力挖掘机，如果你能娴熟地驾驶它们，就能大幅提高开采效率。

本书讲解目前最流行的 Python 爬虫框架 Scrapy，它简单易用、灵活易拓展、文档丰富、开发社区活跃，使用 Scrapy 可以高效地开发网络爬虫应用。本书的读者只需要有 Python 语言基础即可，我们从零基础、逐步由浅入深进行讲解。第 1~8 章讲解 Scrapy 开发的核心基础部分，其中包括：

- 初识 Scrapy
- 编写 Spider
- 使用 Selector 提取数据
- 使用 Item 封装数据
- 使用 Item Pipeline 处理数据
- 使用 Link Extractor 提取链接
- 使用 Exporter 导出数据
- 项目练习

第 9~14 章讲解实际爬虫开发中使用频率最高的一些实用技术，其中包括：

- 下载文件和图片
- 模拟登录
- 爬取动态页面
- 存入数据库
- 使用 HTTP 代理
- 分布式爬取

本书特色

本书的宗旨是以实用和实战为教学目标，主要特色是：

- 所有基础部分的讲解都配有代码示例，而不仅仅是枯燥的文档。
- 案例选材方面以讲解知识点为核心，尽量选择专门供练习爬虫技术的网站（不易变动）或贴近日常生活的网站（京东、知乎、豆瓣、360）进行演示。
- 在讲解某些知识点时，对 Scrapy 源码进行分析，让读者能够"知其然并知其所以然"。

另外，Python 是一门简单易学、功能强大、开发效率极高的语言，近年来在网络爬虫、数据分析、机器学习等领域得到广泛认可。虽然 Python 很容易上手，但想灵活恰当地运用它也并不简单。作者在慕课网（www.imooc.com）上推出了一套《Python 高级进阶实战》课程，可供有需求的读者进行参考：http://coding.imooc.com/class/62.html。

作者还专门针对本书录制了视频课程，感兴趣的读者可登录腾讯网络课堂进行学习，网络课程地址：https://ke.qq.com/course/271952?tuin=c1cd03f4。

致谢

感谢康烁和陈渝老师在清华大学信息研究院工作期间对我在专业方面的耐心指导。
感谢清华大学出版社的王金柱编辑给予我这次写作的机会以及在写作方面的指点。
感谢赵佳音同事认真阅读全书并提出了许多的宝贵建议。
感谢剑超和任怡同学认真审阅全书并对书中代码在多个 Python 版本上进行测试。
感谢女儿刘真，她的笑容化解了写作本书时偶尔的小烦躁。

编　者

2017 年 8 月 8 日

目 录

第 1 章 初识 Scrapy ················ 1
  1.1 网络爬虫是什么 ············· 1
  1.2 Scrapy 简介及安装 ·········· 2
  1.3 编写第一个 Scrapy 爬虫 ····· 3
    1.3.1 项目需求 ··············· 4
    1.3.2 创建项目 ··············· 4
    1.3.3 分析页面 ··············· 5
    1.3.4 实现 Spider ············ 6
    1.3.5 运行爬虫 ··············· 8
  1.4 本章小结 ··················· 11

第 2 章 编写 Spider ··············· 12
  2.1 Scrapy 框架结构及工作原理 ·· 12
  2.2 Request 和 Response 对象 ··· 14
    2.2.1 Request 对象 ············ 15
    2.2.2 Response 对象 ··········· 16
  2.3 Spider 开发流程 ············ 18
    2.3.1 继承 scrapy.Spider ······ 19
    2.3.2 为 Spider 命名 ·········· 20
    2.3.3 设定起始爬取点 ········· 20
    2.3.4 实现页面解析函数 ······· 22
  2.4 本章小结 ··················· 22

第 3 章 使用 Selector 提取数据 ····· 23
  3.1 Selector 对象 ··············· 23

  3.1.1 创建对象 ················· 24
  3.1.2 选中数据 ················· 25
  3.1.3 提取数据 ················· 26
  3.2 Response 内置 Selector ······ 28
  3.3 XPath ······················ 29
    3.3.1 基础语法 ··············· 30
    3.3.2 常用函数 ··············· 35
  3.4 CSS 选择器 ················· 36
  3.5 本章小结 ··················· 40

第 4 章 使用 Item 封装数据 ········ 41
  4.1 Item 和 Field ··············· 42
  4.2 拓展 Item 子类 ············· 44
  4.3 Field 元数据 ················ 44
  4.4 本章小结 ··················· 47

第 5 章 使用 Item Pipeline 处理数据 ·· 48
  5.1 Item Pipeline ················ 48
    5.1.1 实现 Item Pipeline ······· 49
    5.1.2 启用 Item Pipeline ······· 50
  5.2 更多例子 ··················· 51
    5.2.1 过滤重复数据 ··········· 51
    5.2.2 将数据存入
         MongoDB ··············· 54
  5.3 本章小结 ··················· 57

第 6 章 使用 LinkExtractor 提取链接 … 58

- 6.1 使用 LinkExtractor …………… 59
- 6.2 描述提取规则 ………………… 60
- 6.3 本章小结 ……………………… 65

第 7 章 使用 Exporter 导出数据 …… 66

- 7.1 指定如何导出数据 …………… 67
  - 7.1.1 命令行参数 …………… 67
  - 7.1.2 配置文件 ……………… 69
- 7.2 添加导出数据格式 …………… 70
  - 7.2.1 源码参考 ……………… 70
  - 7.2.2 实现 Exporter ………… 72
- 7.3 本章小结 ……………………… 74

第 8 章 项目练习 ……………………… 75

- 8.1 项目需求 ……………………… 77
- 8.2 页面分析 ……………………… 77
- 8.3 编码实现 ……………………… 83
- 8.4 本章小结 ……………………… 88

第 9 章 下载文件和图片 …………… 89

- 9.1 FilesPipeline 和 ImagesPipeline …………………… 89
  - 9.1.1 FilesPipeline 使用说明 ……………………… 90
  - 9.1.2 ImagesPipeline 使用说明 ……………………… 91
- 9.2 项目实战：爬取 matplotlib 例子源码文件 …………… 92
  - 9.2.1 项目需求 ……………… 92
  - 9.2.2 页面分析 ……………… 94
  - 9.2.3 编码实现 ……………… 96
- 9.3 项目实战：下载 360 图片 …… 103
  - 9.3.1 项目需求 ……………… 104
  - 9.3.2 页面分析 ……………… 104
  - 9.3.3 编码实现 ……………… 107
- 9.4 本章小结 ……………………… 109

第 10 章 模拟登录 …………………… 110

- 10.1 登录实质 …………………… 110
- 10.2 Scrapy 模拟登录 …………… 114
  - 10.2.1 使用 FormRequest … 114
  - 10.2.2 实现登录 Spider …… 117
- 10.3 识别验证码 ………………… 119
  - 10.3.1 OCR 识别 …………… 119
  - 10.3.2 网络平台识别 ……… 123
  - 10.3.3 人工识别 …………… 127
- 10.4 Cookie 登录 ………………… 128
  - 10.4.1 获取浏览器 Cookie … 128
  - 10.4.2 CookiesMiddleware 源码分析 …………… 129
  - 10.4.3 实现 BrowserCookies-Middleware ………… 132
  - 10.4.4 爬取知乎个人信息 ………………… 133
- 10.5 本章小结 …………………… 135

第 11 章 爬取动态页面 ……………… 136

- 11.1 Splash 渲染引擎 …………… 140
  - 11.1.1 render.html 端点 …… 141

11.1.2　execute 端点 ……………… 142
11.2　在 Scrapy 中使用 Splash …… 145
11.3　项目实战：爬取 toscrape
　　　中的名人名言 …………………… 146
　　11.3.1　项目需求 …………………… 146
　　11.3.2　页面分析 …………………… 146
　　11.3.3　编码实现 …………………… 147
11.4　项目实战：爬取京东商城
　　　中的书籍信息 …………………… 149
　　11.4.1　项目需求 …………………… 149
　　11.4.2　页面分析 …………………… 149
　　11.4.3　编码实现 …………………… 152
11.5　本章小结 ……………………… 154

第 12 章　存入数据库 ……………… 155

12.1　SQLite ………………………… 156
12.2　MySQL ……………………… 159
12.3　MongoDB …………………… 165
12.4　Redis ………………………… 169
12.5　本章小结 ……………………… 173

第 13 章　使用 HTTP 代理 ………… 174

13.1　HttpProxyMiddleware ……… 175
　　13.1.1　使用简介 …………………… 175
　　13.1.2　源码分析 …………………… 177

13.2　使用多个代理 ………………… 179
13.3　获取免费代理 ………………… 180
13.4　实现随机代理 ………………… 184
13.5　项目实战：爬取豆瓣电影
　　　信息 ……………………………… 187
　　13.5.1　项目需求 …………………… 188
　　13.5.2　页面分析 …………………… 189
　　13.5.3　编码实现 …………………… 194
13.6　本章小结 ……………………… 198

第 14 章　分布式爬取 ……………… 199

14.1　Redis 的使用 ………………… 200
　　14.1.1　安装 Redis ………………… 200
　　14.1.2　Redis 基本命令 …………… 201
14.2　scrapy-redis 源码分析 ……… 206
　　14.2.1　分配爬取任务
　　　　　　部分 …………………………… 207
　　14.2.2　汇总爬取数据
　　　　　　部分 …………………………… 214
14.3　使用 scrapy-redis 进行分
　　　布式爬取 ………………………… 217
　　14.3.1　搭建环境 …………………… 217
　　14.3.2　项目实战 …………………… 218
14.4　本章小结 ……………………… 224

第 1 章

初识 Scrapy

本章首先介绍爬虫的基本概念、工作流程，然后介绍 Scrapy 的安装和网络爬虫项目的实现流程，使读者对网络爬虫有一个大致的了解，并且建立起网络爬虫的编写思路。本章重点讲解以下内容：

- 网络爬虫及爬虫的工作流程。
- Scrapy 的介绍与安装。
- 网络爬虫编写步骤。

1.1 网络爬虫是什么

网络爬虫是指在互联网上自动爬取网站内容信息的程序，也被称作网络蜘蛛或网络机器人。大型的爬虫程序被广泛应用于搜索引擎、数据挖掘等领域，个人用户或企业也可以利用爬虫收集对自身有价值的数据。举一个简单的例子，假设你在本地新开了一家以外卖生意为主的餐馆，现在要给菜品定价，此时便可以开发一个爬虫程序，在美团、饿了么、百度外卖这些外卖网站爬取大量其他餐馆的菜品价格作为参考，以指导定价。

一个网络爬虫程序的基本执行流程可以总结为以下循环：

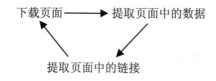

1. 下载页面

一个网页的内容本质上就是一个 HTML 文本，爬取一个网页内容之前，首先要根据网页的 URL 下载网页。

2. 提取页面中的数据

当一个网页（HTML）下载完成后，对页面中的内容进行分析，并提取出我们感兴趣的数据，提取到的数据可以以多种形式保存起来，比如将数据以某种格式（CSV、JSON）写入文件中，或存储到数据库（MySQL、MongoDB）中。

3. 提取页面中的链接

通常，我们想要获取的数据并不只在一个页面中，而是分布在多个页面中，这些页面彼此联系，一个页面中可能包含一个或多个到其他页面的链接，提取完当前页面中的数据后，还要把页面中的某些链接也提取出来，然后对链接页面进行爬取（循环 1-3 步骤）。

设计爬虫程序时，还要考虑防止重复爬取相同页面（URL 去重）、网页搜索策略（深度优先或广度优先等）、爬虫访问边界限定等一系列问题。

从头开发一个爬虫程序是一项烦琐的工作，为了避免因制造轮子而消耗大量时间，在实际应用中我们可以选择使用一些优秀的爬虫框架，使用框架可以降低开发成本，提高程序质量，让我们能够专注于业务逻辑（爬取有价值的数据）。接下来，本书就带你学习目前非常流行的开源爬虫框架 Scrapy。

1.2　Scrapy 简介及安装

Scrapy 是一个使用 Python 语言（基于 Twisted 框架）编写的开源网络爬虫框架，目前由 Scrapinghub Ltd 维护。Scrapy 简单易用、灵活易拓展、开发社区活跃，并且是跨平台的。在 Linux、MaxOS 以及 Windows 平台都可以使用。Scrapy 应用程序也使用 Python 进行开发，目前可以支持 Python 2.7 以及 Python 3.4+版本。

在任意操作系统下，可以使用 pip 安装 Scrapy，例如：

```
$ pip install scrapy
```

为确认 Scrapy 已安装成功，首先在 Python 中测试能否导入 Scrapy 模块：

```
>>> import scrapy
>>> scrapy.version_info
(1, 3, 3)
```

然后，在 shell 中测试能否执行 Scrapy 这条命令：

```
$ scrapy
Scrapy 1.3.3 - no active project

Usage:
  scrapy  [options] [args]

Available commands:
  bench          Run quick benchmark test
  commands
  fetch          Fetch a URL using the Scrapy downloader
  genspider      Generate new spider using pre-defined templates
  runspider      Run a self-contained spider (without creating a project)
  settings       Get settings values
  shell          Interactive scraping console
  startproject   Create new project
  version        Print Scrapy version
  view           Open URL in browser, as seen by Scrapy

  [ more ]       More commands available when run from project directory

Use "scrapy  -h" to see more info about a command
```

通过了以上两项检测，说明 Scrapy 安装成功了。如上所示，我们安装的是当前最新版本 1.3.3。

1.3 编写第一个 Scrapy 爬虫

为了帮助大家建立对 Scrapy 框架的初步印象，我们使用它完成一个简单的爬虫项目。

1.3.1 项目需求

在专门供爬虫初学者训练爬虫技术的网站（http://books.toscrape.com）上爬取书籍信息，如图 1-1 所示。

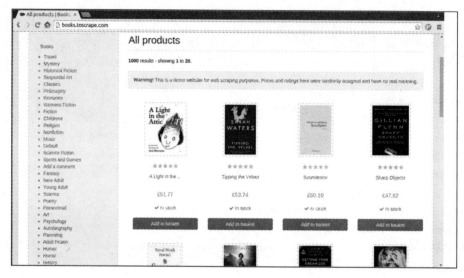

图 1-1

该网站中，这样的书籍列表页面一共有 50 页，每页有 20 本书，第一个例子应尽量简单。我们下面仅爬取所有图书（1000 本）的书名和价格信息。

1.3.2 创建项目

首先，我们要创建一个 Scrapy 项目，在 shell 中使用 scrapy startproject 命令：

```
$ scrapy startproject example
New Scrapy project 'example', using template directory
'/usr/local/lib/python3.4/dist-packages/scrapy/templates/project', created in:
    /home/liushuo/book/example

You can start your first spider with:
    cd example
    scrapy genspider example example.com
```

创建好一个名为 example 的项目后，可使用 tree 命令查看项目目录下的文件，显示如下：

```
$ tree example
example/
├── example
│   ├── __init__.py
│   ├── items.py
│   ├── middlewares.py
│   ├── pipelines.py
│   ├── settings.py
│   └── spiders
│       └── __init__.py
└── scrapy.cfg
```

随着后面逐步深入学习，大家会了解这些文件的用途，此处不做解释。

1.3.3 分析页面

编写爬虫程序之前，首先需要对待爬取的页面进行分析，主流的浏览器中都带有分析页面的工具或插件，这里我们选用 Chrome 浏览器的开发者工具（Tools→Developer tools）分析页面。

1. 数据信息

在 Chrome 浏览器中打开页面 http://books.toscrape.com，选中其中任意一本书并右击，然后选择"审查元素"，查看其 HTML 代码，如图 1-2 所示。

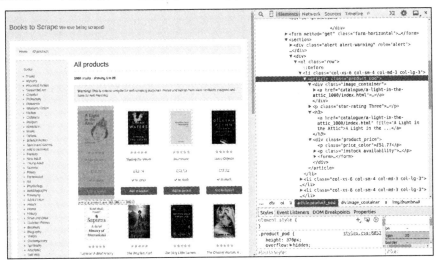

图 1-2

可以看到，每一本书的信息包裹在<article class="product_pod">元素中：书名信息在其下 h3 > a 元素的 title 属性中，如<a href="catalogue/a-light-in-the-attic_1000/index.html" title="A Light in the Attic">A Light in the ...</a>；书价信息在其下<p class="price_color">元素的文本中，如<p class="price_color">£51.77</p>。

2. 链接信息

图 1-3 所示为第一页书籍列表页面，可以通过单击 next 按钮访问下一页，选中页面下方的 next 按钮并右击，然后选择"审查元素"，查看其 HTML 代码，如图 1-3 所示。

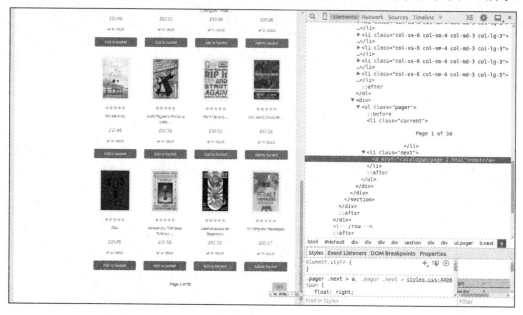

图 1-3

可以发现，下一页的 URL 在 ul.pager > li.next > a 元素的 href 属性中，是一个相对 URL 地址，如<li class="next"><a href="catalogue/page-2.html">next</a></li>。

1.3.4 实现 Spider

分析完页面后，接下来编写爬虫。在 Scrapy 中编写一个爬虫，即实现一个 scrapy.Spider 的子类。

实现爬虫的 Python 文件应位于 exmaple/spiders 目录下，在该目录下创建新文件 book_spider.py。然后，在 book_spider.py 中实现爬虫 BooksSpider，代码如下：

```python
# -*- coding: utf-8 -*-
import scrapy

class BooksSpider(scrapy.Spider):
    # 每一个爬虫的唯一标识
    name = "books"

    # 定义爬虫爬取的起始点,起始点可以是多个,这里只有一个
    start_urls = ['http://books.toscrape.com/']

    def parse(self, response):
        # 提取数据
        # 每一本书的信息在<article class="product_pod">中,我们使用
        # css()方法找到所有这样的 article 元素,并依次迭代
        for book in response.css('article.product_pod'):
            # 书名信息在 article > h3 > a 元素的 title 属性里
            # 例如: <a title="A Light in the Attic">A Light in the ...</a>
            name = book.xpath('./h3/a/@title').extract_first()

            # 书价信息在 <p class="price_color">的 TEXT 中。
            # 例如: <p class="price_color">£51.77</p>
            price = book.css('p.price_color::text').extract_first()
            yield {
                'name': name,
                'price': price,
            }

        # 提取链接
        # 下一页的 url 在 ul.pager > li.next > a 里面
        # 例如: <li class="next"><a href="catalogue/page-2.html">next</a></li>
        next_url = response.css('ul.pager li.next a::attr(href)').extract_first()
        if next_url:
            # 如果找到下一页的 URL,得到绝对路径,构造新的 Request 对象
            next_url = response.urljoin(next_url)
            yield scrapy.Request(next_url, callback=self.parse)
```

如果上述代码中有看不懂的部分,大家不必担心,更多详细内容会在后面章节学习,这里只要先对实现一个爬虫有个整体印象即可。

下面对 BooksSpider 的实现做简单说明。

- name 属性

  一个 Scrapy 项目中可能有多个爬虫，每个爬虫的 name 属性是其自身的唯一标识，在一个项目中不能有同名的爬虫，本例中的爬虫取名为'books'。

- start_urls 属性

  一个爬虫总要从某个（或某些）页面开始爬取，我们称这样的页面为起始爬取点，start_urls 属性用来设置一个爬虫的起始爬取点。在本例中只有一个起始爬取点'http://books.toscrape.com'。

- parse 方法

  当一个页面下载完成后，Scrapy 引擎会回调一个我们指定的页面解析函数（默认为 parse 方法）解析页面。一个页面解析函数通常需要完成以下两个任务：
  - 提取页面中的数据（使用 XPath 或 CSS 选择器）。
  - 提取页面中的链接，并产生对链接页面的下载请求。

页面解析函数通常被实现成一个生成器函数，每一项从页面中提取的数据以及每一个对链接页面的下载请求都由 yield 语句提交给 Scrapy 引擎。

1.3.5 运行爬虫

完成代码后，运行爬虫爬取数据，在 shell 中执行 scrapy crawl <SPIDER_NAME> 命令运行爬虫'books'，并将爬取的数据存储到 csv 文件中：

```
$ scrapy crawl books -o books.csv
2016-12-27 15:19:53 [scrapy] INFO: Scrapy 1.3.3 started (bot: example)
2016-12-27 15:19:53 [scrapy] INFO: INFO: Overridden settings: {'BOT_NAME': 'example', 'SPIDER_MODULES': ['example.spiders'], 'ROBOTSTXT_OBEY': True, 'NEWSPIDER_MODULE': 'example.spiders'}
2016-12-27 15:19:53 [scrapy] INFO: Enabled extensions:
['scrapy.extensions.telnet.TelnetConsole',
 'scrapy.extensions.corestats.CoreStats',
 'scrapy.extensions.feedexport.FeedExporter',
 'scrapy.extensions.logstats.LogStats']
2016-12-27 15:19:53 [scrapy] INFO: Enabled downloader middlewares:
['scrapy.downloadermiddlewares.robotstxt.RobotsTxtMiddleware',
 'scrapy.downloadermiddlewares.httpauth.HttpAuthMiddleware',
 'scrapy.downloadermiddlewares.downloadtimeout.DownloadTimeoutMiddleware',
 'scrapy.downloadermiddlewares.defaultheaders.DefaultHeadersMiddleware',
 'scrapy.downloadermiddlewares.useragent.UserAgentMiddleware',
 'scrapy.downloadermiddlewares.retry.RetryMiddleware',
```

```
    'scrapy.downloadermiddlewares.redirect.MetaRefreshMiddleware',
    'scrapy.downloadermiddlewares.httpcompression.HttpCompressionMiddleware',
    'scrapy.downloadermiddlewares.redirect.RedirectMiddleware',
    'scrapy.downloadermiddlewares.cookies.CookiesMiddleware',
    'scrapy.downloadermiddlewares.chunked.ChunkedTransferMiddleware',
    'scrapy.downloadermiddlewares.stats.DownloaderStats']
2016-12-27 15:19:53 [scrapy] INFO: Enabled spider middlewares:
['scrapy.spidermiddlewares.httperror.HttpErrorMiddleware',
    'scrapy.spidermiddlewares.offsite.OffsiteMiddleware',
    'scrapy.spidermiddlewares.referer.RefererMiddleware',
    'scrapy.spidermiddlewares.urllength.UrlLengthMiddleware',
    'scrapy.spidermiddlewares.depth.DepthMiddleware']
2016-12-27 15:19:53 [scrapy] INFO: Enabled item pipelines:
[]
2016-12-27 15:19:53 [scrapy] INFO: Spider opened
2016-12-27 15:19:53 [scrapy] INFO: Crawled 0 pages (at 0 pages/min), scraped 0 items (at 0 items/min)
2016-12-27 15:19:53 [scrapy] DEBUG: Telnet console listening on 127.0.0.1:6023
2016-12-27 15:20:01 [scrapy] DEBUG: Crawled (404)    (referer: None)
2016-12-27 15:20:02 [scrapy] DEBUG: Crawled (200)    (referer: None)
2016-12-27 15:20:02 [scrapy] DEBUG: Scraped from <200 http://books.toscrape.com/>
    {'name': 'A Light in the Attic', 'price': '£51.77'}
2016-12-27 15:20:02 [scrapy] DEBUG: Scraped from <200 http://books.toscrape.com/>
    {'name': 'Tipping the Velvet', 'price': '£53.74'}
2016-12-27 15:20:02 [scrapy] DEBUG: Scraped from <200 http://books.toscrape.com/>
    {'name': 'Soumission', 'price': '£50.10'}

... <省略中间部分输出> ...

2016-12-27 15:21:30 [scrapy] DEBUG: Scraped from <200 http://books.toscrape.com/catalogue/page-50.html>
    {'name': '1,000 Places to See Before You Die', 'price': '£26.08'}
2016-12-27 15:21:30 [scrapy] INFO: Closing spider (finished)
2016-12-27 15:21:30 [scrapy] INFO: Stored csv feed (1000 items) in: books.csv
2016-12-27 15:21:30 [scrapy] INFO: Dumping Scrapy stats:
{'downloader/request_bytes': 14957,
 'downloader/request_count': 51,
 'downloader/request_method_count/GET': 51,
```

```
 'downloader/response_bytes': 299924,
 'downloader/response_count': 51,
 'downloader/response_status_count/200': 50,
 'downloader/response_status_count/404': 1,
 'finish_reason': 'finished',
 'finish_time': datetime.datetime(2016, 12, 27, 7, 21, 30, 10396),
 'item_scraped_count': 1000,
 'log_count/DEBUG': 1052,
 'log_count/INFO': 9,
 'request_depth_max': 49,
 'response_received_count': 51,
 'scheduler/dequeued': 50,
 'scheduler/dequeued/memory': 50,
 'scheduler/enqueued': 50,
 'scheduler/enqueued/memory': 50,
 'start_time': datetime.datetime(2016, 12, 27, 7, 19, 53, 194334)}
2016-12-27 15:21:30 [scrapy] INFO: Spider closed (finished)
```

等待爬虫运行结束后，在 books.csv 文件中查看爬取到的数据，代码如下：

```
$ sed -n '2,$p' books.csv | cat -n        # 不显示第一行的 csv 头部
     1  A Light in the Attic,£51.77
     2  Tipping the Velvet,£53.74
     3  Soumission,£50.10
     4  Sharp Objects,£47.82
     5  Sapiens: A Brief History of Humankind,£54.23
     6  The Requiem Red,£22.65
     7  The Dirty Little Secrets of Getting Your Dream Job,£33.34

     ...<省略中间部分输出>...

   995  Beyond Good and Evil,£43.38
   996  Alice in Wonderland (Alice's Adventures in Wonderland #1),£55.53
   997  "Ajin: Demi-Human, Volume 1 (Ajin: Demi-Human #1)",£57.06
   998  A Spy's Devotion (The Regency Spies of London #1),£16.97
   999  1st to Die (Women's Murder Club #1),£53.98
  1000  "1,000 Places to See Before You Die",£26.08
```

从上述数据可以看出，我们成功地爬取到了 1000 本书的书名和价格信息（50 页，每页 20 项）。

1.4 本章小结

本章是开始 Scrapy 爬虫之旅的第 1 章，先带大家了解了什么是网络爬虫，然后对 Scrapy 爬虫框架做了简单介绍，最后以一个简单的爬虫项目让大家对开发 Scrapy 爬虫有了初步的印象。在接下来的章节中，我们将深入学习开发 Scrapy 爬虫的核心基础内容。

第 2 章

编写 Spider

从本章开始介绍 Scrapy 爬虫的基础知识部分，我们先从 Scrapy 爬虫程序中最核心的组件 Spider 讲起，在学习编写 Spider 之前，需要一些知识作为铺垫。本章重点讲解以下内容：

- Scrapy 框架结构及工作原理。
- Request 对象和 Response 对象。

2.1　Scrapy 框架结构及工作原理

图 2-1 展示了 Scrapy 框架的组成结构，并从数据流的角度揭示 Scrapy 的工作原理。

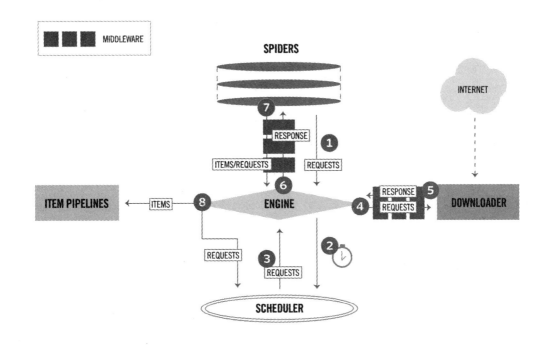

图 2-1

首先，简单了解一下 Scrapy 框架中的各个组件，如表 2-1 所示。

表 2-1

组件	描述	类型
ENGINE	引擎，框架的核心，其他所有组件在其控制下协同工作	内部组件
SCHEDULER	调度器，负责对 SPIDER 提交的下载请求进行调度	内部组件
DOWNLOADER	下载器，负责下载页面（发送 HTTP 请求/接收 HTTP 响应）	内部组件
SPIDER	爬虫，负责提取页面中的数据，并产生对新页面的下载请求	用户实现
MIDDLEWARE	中间件，负责对 Request 对象和 Response 对象进行处理	可选组件
ITEM PIPELINE	数据管道，负责对爬取到的数据进行处理	可选组件

对于用户来说，Spider 是最核心的组件，Scrapy 爬虫开发是围绕实现 Spider 展开的。接下来，看一下在框架中的数据流，有表 2-2 所示的 3 种对象。

表 2-2

对象	描述
REQUEST	Scrapy 中的 HTTP 请求对象
RESPONSE	Scrapy 中的 HTTP 响应对象
ITEM	从页面中爬取的一项数据

Request 和 Response 是 HTTP 协议中的术语，即 HTTP 请求和 HTTP 响应，Scrapy 框架中定义了相应的 Request 和 Response 类，这里的 Item 代表 Spider 从页面中爬取的一项数据。

最后，我们来说明以上几种对象在框架中的流动过程。

- 当 SPIDER 要爬取某 URL 地址的页面时，需使用该 URL 构造一个 Request 对象，提交给 ENGINE（图 2-1 中的 1）。
- Request 对象随后进入 SCHEDULER 按某种算法进行排队，之后的某个时刻 SCHEDULER 将其出队，送往 DOWNLOADER（图 2-1 中的 2、3、4）。
- DOWNLOADER 根据 Request 对象中的 URL 地址发送一次 HTTP 请求到网站服务器，之后用服务器返回的 HTTP 响应构造出一个 Response 对象，其中包含页面的 HTML 文本（图 2-1 中的 5）。
- Response 对象最终会被递送给 SPIDER 的页面解析函数（构造 Request 对象时指定）进行处理，页面解析函数从页面中提取数据，封装成 Item 后提交给 ENGINE，Item 之后被送往 ITEM PIPELINES 进行处理，最终可能由 EXPORTER（图 2-1 中没有显示）以某种数据格式写入文件（csv, json）；另一方面，页面解析函数还从页面中提取链接（URL），构造出新的 Request 对象提交给 ENGINE（图 2-1 中的 6、7、8）。

理解了框架中的数据流，也就理解了 Scrapy 爬虫的工作原理。如果把框架中的组件比作人体的各个器官，Request 和 Response 对象便是血液，Item 则是代谢产物。

2.2 Request 和 Response 对象

通过上述讲解，大家已经了解了 Request 和 Response 对象在 Scrapy 框架中的重要性，下面详细介绍这两个对象。

2.2.1 Request 对象

Request 对象用来描述一个 HTTP 请求，下面是其构造器方法的参数列表：

```
Request(url[, callback, method='GET', headers, body, cookies, meta,
        encoding='utf-8', priority=0, dont_filter=False, errback])
```

下面依次介绍这些参数。

- url（必选）
  请求页面的 url 地址，bytes 或 str 类型，如'http://www.python.org/doc'。
- callback
  页面解析函数，Callable 类型，Request 对象请求的页面下载完成后，由该参数指定的页面解析函数被调用。如果未传递该参数，默认调用 Spider 的 parse 方法。
- method
  HTTP 请求的方法，默认为'GET'。
- headers
  HTTP 请求的头部字典，dict 类型，例如 {'Accept': 'text/html', 'User-Agent': 'Mozilla/5.0'}。如果其中某项的值为 None，就表示不发送该项 HTTP 头部，例如 {'Cookie': None}，禁止发送 Cookie。
- body
  HTTP 请求的正文，bytes 或 str 类型。
- cookies
  Cookie 信息字典，dict 类型，例如 {'currency': 'USD', 'country': 'UY'}。
- meta
  Request 的元数据字典，dict 类型，用于给框架中其他组件传递信息，比如中间件 Item Pipeline。其他组件可以使用 Request 对象的 meta 属性访问该元数据字典（request.meta），也用于给响应处理函数传递信息，详见 Response 的 meta 属性。
- encoding
  url 和 body 参数的编码默认为'utf-8'。如果传入的 url 或 body 参数是 str 类型，就使用该参数进行编码。
- priority
  请求的优先级默认值为 0，优先级高的请求优先下载。
- dont_filter
  默认情况下（dont_filter=False），对同一个 url 地址多次提交下载请求，后面的请求会被去重过滤器过滤（避免重复下载）。如果将该参数置为 True，可以使请求避免被过滤，强制下载。例如，在多次爬取一个内容随时间而变化的页面时（每次使用相同的url），可以将该参数置为True。
- errback
请求出现异常或者出现HTTP错误时（如404页面不存在）的回调函数。

虽然参数很多，但除了url参数外，其他都带有默认值。在构造Request对象时，通常我们只需传递一个url参数或再加一个callback参数，其他使用默认值即可，代码如下：

```
>>> import scrapy
>>> request = scrapy.Request('http://books.toscrape.com/')
>>> request2 = scrapy.Request('http://quotes.toscrape.com/', callback=self.parseItem)
```

在实际应用中，我们几乎只调用Request的构造器创建对象，但也可以根据需求访问Request对象的属性，常用的有以下几个：

- url
- method
- headers
- body
- meta

这些属性和构造器参数相对应，这里不再重复解释。

2.2.2 Response对象

Response对象用来描述一个HTTP响应，Response只是一个基类，根据响应内容的不同有如下子类：

- TextResponse
- HtmlResponse
- XmlResponse

当一个页面下载完成时，下载器依据HTTP响应头部中的Content-Type信息创建某个Response的子类对象。我们通常爬取的网页，其内容是HTML文本，创建的便是HtmlResponse对象，其中HtmlResponse和XmlResponse是TextResponse的子类。实际上，这3个子类只有细微的差别，这里以HtmlResponse为例进行讲解。

下面介绍HtmlResponse对象的属性及方法。

- url
  HTTP 响应的 url 地址，str 类型。
- status
  HTTP 响应的状态码，int 类型，例如 200, 404。
- headers
  HTTP 响应的头部，类字典类型，可以调用 get 或 getlist 方法对其进行访问，例如：

```
response.headers.get('Content-Type')
response.headers.getlist('Set-Cookie')
```

- body
  HTTP 响应正文，bytes 类型。
- text
  文本形式的 HTTP 响应正文，str 类型，它是由 response.body 使用 response.encoding 解码得到的，即

```
reponse.text = response.body.decode(response.encoding)
```

- encoding
  HTTP 响应正文的编码，它的值可能是从 HTTP 响应头部或正文中解析出来的。
- request
  产生该 HTTP 响应的 Request 对象。
- meta
  即 response.request.meta，在构造 Request 对象时，可将要传递给响应处理函数的信息通过 meta 参数传入；响应处理函数处理响应时，通过 response.meta 将信息取出。
- selector
  Selector 对象用于在 Response 中提取数据（选择器相关话题在后面章节详细讲解）。
- xpath(query)
  使用 XPath 选择器在 Response 中提取数据，实际上它是 response.selector.xpath 方法的快捷方式（选择器相关话题在后面章节详细讲解）。
- css(query)
  使用 CSS 选择器在 Response 中提取数据，实际上它是 response.selector.css 方法的快捷方式（选择器相关话题在后面章节详细讲解）。
- urljoin（url）
  用于构造绝对 url。当传入的 url 参数是一个相对地址时，根据 response.url 计算出相应的绝对 url。例如，response.url 为 http://www.example.com/a，url 为 b/index.html，调用 response.urljoin(url) 的结果为 http://www.example.com/a/b/index.html。

虽然 HtmlResponse 对象有很多属性，但最常用的是以下的 3 个方法：

- xpath(query)
- css(query)
- urljoin(url)

前两个方法用于提取数据，后一个方法用于构造绝对 url。

2.3 Spider 开发流程

有了前面知识的铺垫，现在回到本章的主题"编写 Spider"。实现一个 Spider 子类的过程很像是完成一系列填空题，Scrapy 框架提出以下问题让用户在 Spider 子类中作答：

- 爬虫从哪个或哪些页面开始爬取？
- 对于一个已下载的页面，提取其中的哪些数据？
- 爬取完当前页面后，接下来爬取哪个或哪些页面？

上面问题的答案包含了一个爬虫最重要的逻辑，回答了这些问题，一个爬虫也就开发出来了。

接下来，我们以上一章 exmaple 项目中的 BooksSpider 为例，讲解一个 Spider 的开发流程。为方便阅读，再次给出 BooksSpider 的代码：

```
# -*- coding: utf-8 -*-
import scrapy

class BooksSpider(scrapy.Spider):
    # 每一个爬虫的唯一标识
    name = "books"

    # 定义爬虫爬取的起始点，起始点可以是多个，我们这里是一个
    start_urls = ['http://books.toscrape.com/']

    def parse(self, response):
        # 提取数据
        # 每一本书的信息是在<article class="product_pod">中，我们使用
        # css()方法找到所有这样的 article 元素，并依次迭代
        for book in response.css('article.product_pod'):
            # 书名信息在 article > h3 > a 元素的 title 属性里
```

```
        # 例如：<a title="A Light in the Attic">A Light in the ...</a>
        name = book.xpath('./h3/a/@title').extract_first()

        # 书价信息在 <p class="price_color">的 TEXT 中。
        # 例如：<p class="price_color">£51.77</p>
        price = book.css('p.price_color::text').extract_first()
        yield {
            'name': name,
            'price': price,
        }

    # 提取链接
    # 下一页的 url 在 ul.pager > li.next > a 里面
    # 例如：<li class="next"><a href="catalogue/page-2.html">next</a></li>
    next_url = response.css('ul.pager li.next a::attr(href)').extract_first()
    if next_url:
        # 如果找到下一页的 url, 得到绝对路径, 构造新的 Request 对象
        next_url = response.urljoin(next_url)
        yield scrapy.Request(next_url, callback=self.parse)
```

实现一个 Spider 只需要完成下面 4 个步骤：

步骤 01 继承 scrapy.Spider。
步骤 02 为 Spider 取名。
步骤 03 设定起始爬取点。
步骤 04 实现页面解析函数。

2.3.1 继承 scrapy.Spider

Scrapy 框架提供了一个 Spider 基类，我们编写的 Spider 需要继承它：

```
import scrapy
class BooksSpider(scrapy.Spider):
    ...
```

这个 Spider 基类实现了以下内容：

- 供 Scrapy 引擎调用的接口，例如用来创建 Spider 实例的类方法 from_crawler。
- 供用户使用的实用工具函数，例如可以调用 log 方法将调试信息输出到日志。
- 供用户访问的属性，例如可以通过 settings 属性访问配置文件中的配置。

实际上，在初学 Scrapy 时，不必关心 Spider 基类的这些细节，未来有需求时再去查阅文档即可。

2.3.2 为 Spider 命名

在一个 Scrapy 项目中可以实现多个 Spider，每个 Spider 需要有一个能够区分彼此的唯一标识，Spider 的类属性 name 便是这个唯一标识。

```
class BooksSpider(scrapy.Spider):
    name = "books"
    ...
```

执行 scrapy crawl 命令时就用到了这个标识，告诉 Scrapy 使用哪个 Spider 进行爬取。

2.3.3 设定起始爬取点

Spider 必然要从某个或某些页面开始爬取，我们称这些页面为起始爬取点，可以通过类属性 start_urls 来设定起始爬取点：

```
class BooksSpider(scrapy.Spider):
    ...
    start_urls = ['http://books.toscrape.com/']
    ...
```

start_urls 通常被实现成一个列表，其中放入所有起始爬取点的 url（例子中只有一个起始点）。看到这里，大家可能会想，请求页面下载不是一定要提交 Request 对象么？而我们仅定义了 url 列表，是谁暗中构造并提交了相应的 Request 对象呢？通过阅读 Spider 基类的源码可以找到答案，相关代码如下：

```
class Spider(object_ref):
    ...
    def start_requests(self):
        for url in self.start_urls:
            yield self.make_requests_from_url(url)

    def make_requests_from_url(self, url):
        return Request(url, dont_filter=True)

    def parse(self, response):
        raise NotImplementedError
    ...
```

从代码中可以看出，Spider 基类的 start_requests 方法帮助我们构造并提交了 Request 对象，对其中的原理做如下解释：

- 实际上，对于起始爬取点的下载请求是由 Scrapy 引擎调用 Spider 对象的 start_requests 方法提交的，由于 BooksSpider 类没有实现 start_requests 方法，因此引擎会调用 Spider 基类的 start_requests 方法。
- 在 start_requests 方法中，self.start_urls 便是我们定义的起始爬取点列表（通过实例访问类属性），对其进行迭代，用迭代出的每个 url 作为参数调用 make_requests_from_url 方法。
- 在 make_requests_from_url 方法中，我们找到了真正构造 Reqeust 对象的代码，仅使用 url 和 dont_filter 参数构造 Request 对象。
- 由于构造 Request 对象时并没有传递 callback 参数来指定页面解析函数，因此默认将 parse 方法作为页面解析函数。此时 BooksSpider 必须实现 parse 方法，否则就会调用 Spider 基类的 parse 方法，从而抛出 NotImplementedError 异常（可以看作基类定义了一个抽象接口）。
- 起始爬取点可能有多个，start_requests 方法需要返回一个可迭代对象（列表、生成器等），其中每一个元素是一个 Request 对象。这里，start_requests 方法被实现成一个生成器函数（生成器对象是可迭代的），每次由 yield 语句返回一个 Request 对象。

由于起始爬取点的下载请求是由引擎调用 Spider 对象的 start_requests 方法产生的，因此我们也可以在 BooksSpider 中实现 start_requests 方法（覆盖基类 Spider 的 start_requests 方法），直接构造并提交起始爬取点的 Request 对象。在某些场景下使用这种方式更加灵活，例如有时想为 Request 添加特定的 HTTP 请求头部，或想为 Request 指定特定的页面解析函数。

以下是通过实现 start_requests 方法定义起始爬取点的示例代码（改写 BooksSpider）：

```python
class BooksSpider(scrapy.Spider):

    # start_urls = ['http://books.toscrape.com/']

    # 实现 start_requests 方法, 替代 start_urls 类属性
    def start_requests(self):
        yield scrapy.Request('http://books.toscrape.com/',
                             callback=self.parse_book,
                             headers={'User-Agent': 'Mozilla/5.0'},
                             dont_filter=True)
    # 改用 parse_book 作为回调函数
```

```
def parse_book(response):
    ...
```

到此,我们介绍完了为爬虫设定起始爬取点的两种方式:

- 定义 start_urls 属性。
- 实现 start_requests 方法。

2.3.4 实现页面解析函数

页面解析函数也就是构造 Request 对象时通过 callback 参数指定的回调函数(或默认的 parse 方法)。页面解析函数是实现 Spider 中最核心的部分,它需要完成以下两项工作:

- 使用选择器提取页面中的数据,将数据封装后(Item 或字典)提交给 Scrapy 引擎。
- 使用选择器或 LinkExtractor 提取页面中的链接,用其构造新的 Request 对象并提交给 Scrapy 引擎(下载链接页面)。

一个页面中可能包含多项数据以及多个链接,因此页面解析函数被要求返回一个可迭代对象(通常被实现成一个生成器函数),每次迭代返回一项数据(Item 或字典)或一个 Request 对象。

关于如何提取数据、封装数据、提取链接等话题,我们在接下来的章节继续学习。

2.4 本章小结

本章先讲解了 Scrapy 的框架结构以及工作原理,然后又介绍了 Scrapy 中与页面下载相关的两个核心对象 Request 和 Response,有了这些知识的铺垫,最后我们讲解了实现一个 Spider 的开发流程。关于编写 Spider 时涉及的一些技术细节在后面章节继续讨论。

第 3 章

使用 Selector 提取数据

在第 2 章中，我们讲解了 Spider 开发的大体流程，让大家对 Scrapy 爬虫开发有了更加清晰的理解，接下来继续讲解编写爬虫时使用的一些具体技术，从页面中提取数据是 Spider 最重要的工作之一，这一章我们来学习相关内容。

3.1 Selector 对象

从页面中提取数据的核心技术是 HTTP 文本解析，在 Python 中常用以下模块处理此类问题：

- BeautifulSoup
  BeautifulSoup 是非常流行的 HTTP 解析库，API 简洁易用，但解析速度较慢。
- lxml
  lxml 是一套由 C 语言编写的 xml 解析库（libxml2），解析速度更快，API 相对复杂。

Scrapy 综合上述两者优点实现了 Selector 类，它是基于 lxml 库构建的，并简化了 API 接口。在 Scrapy 中使用 Selector 对象提取页面中的数据，使用时先通过 XPath 或

CSS 选择器选中页面中要提取的数据,然后进行提取。

下面详细介绍 Selector 对象的使用。

3.1.1 创建对象

Selector 类的实现位于 scrapy.selector 模块,创建 Selector 对象时,可将页面的 HTML 文档字符串传递给 Selector 构造器方法的 text 参数:

```
>>> from scrapy.selector import Selector
>>> text = '''
... <html>
...     <body>
...         <h1>Hello World</h1>
...         <h1>Hello Scrapy</h1>
...         <b>Hello python</b>
...         <ul>
...             <li>C++</li>
...             <li>Java</li>
...             <li>Python</li>
...         </ul>
...     </body>
... </html>
... '''
>>> selector = Selector(text=text)
<Selector xpath=None data='<html>\n        <body>\n            <h1>He'>
```

也可以使用一个 Response 对象构造 Selector 对象,将其传递给 Selector 构造器方法的 response 参数:

```
>>> from scrapy.selector import Selector
>>> from scrapy.http import HtmlResponse
>>> body = '''
... <html>
...     <body>
...         <h1>Hello World</h1>
...         <h1>Hello Scrapy</h1>
...         <b>Hello python</b>
...         <ul>
```

```
...                    <li>C++</li>
...                    <li>Java</li>
...                    <li>Python</li>
...                </ul>
...            </body>
... </html>
... '''
...
>>> response = HtmlResponse(url='http://www.example.com', body=body, encoding='utf8')
>>> selector = Selector(response=response)
>>> selector
<Selector xpath=None data='<html>\n        <body>\n            <h1>He'>
```

3.1.2 选中数据

调用 Selector 对象的 xpath 方法或 css 方法（传入 XPath 或 CSS 选择器表达式），可以选中文档中的某个或某些部分：

```
>>> selector_list = selector.xpath('//h1')      # 选中文档中所有的<h1>
>>> selector_list                               # 其中包含两个<h1>对应的 Selector 对象
[<Selector xpath='.//h1' data='<h1>Hello World</h1>'>,
 <Selector xpath='.//h1' data='<h1>Hello Scrapy</h1>'>]
```

xpath 和 css 方法返回一个 SelectorList 对象，其中包含每个被选中部分对应的 Selector 对象，SelectorList 支持列表接口，可使用 for 语句迭代访问其中的每一个 Selector 对象：

```
>>> for sel in selector_list:
...     print(sel.xpath('./text()'))
...
[<Selector xpath='./text()' data='Hello World'>]
[<Selector xpath='./text()' data='Hello Scrapy'>]
```

SelectorList 对象也有 xpath 和 css 方法，调用它们的行为是：以接收到的参数分别调用其中每一个 Selector 对象的 xpath 或 css 方法，并将所有结果收集到一个新的 SelectorList 对象返回给用户。请看下面的示例：

```
>>> selector_list.xpath('./text()')
[<Selector xpath='./text()' data='Hello World'>,
 <Selector xpath='./text()' data='Hello Scrapy'>]
```

```
>>> selector.xpath('.//ul').css('li').xpath('./text()')
[<Selector xpath='./text()' data='C++'>,
 <Selector xpath='./text()' data='Java'>,
 <Selector xpath='./text()' data='Python'>]
```

3.1.3 提取数据

调用 Selector 或 SelectorLis 对象的以下方法可将选中的内容提取：

- extract()
- re()
- extract_first()   (SelectorList 专有)
- re_first()   (SelectorList 专有)

首先来看 extract 方法，调用 Selector 对象的 extract 方法将返回选中内容的 Unicode 字符串：

```
>>> sl = selector.xpath('.//li')
>>> sl
[<Selector xpath='.//li' data='<li>C++</li>'>,
 <Selector xpath='.//li' data='<li>Java</li>'>,
 <Selector xpath='.//li' data='<li>Python</li>'>]
>>> sl[0].extract()
'<li>C++</li>'
>>> sl = selector.xpath('.//li/text()')
>>> sl
[<Selector xpath='.//li/text()' data='C++'>,
 <Selector xpath='.//li/text()' data='Java'>,
 <Selector xpath='.//li/text()' data='Python'>]
>>> sl[1].extract()
'Java'
```

与 SelectorList 对象的 xpath 和 css 方法类似，SelectorList 对象的 extract 方法内部会调用其中每个 Selector 对象的 extract 方法，并把所有结果收集到一个列表返回给用户：

```
>>> sl = selector.xpath('.//li/text()')
>>> sl
[<Selector xpath='.//li/text()' data='C++'>,
 <Selector xpath='.//li/text()' data='Java'>,
 <Selector xpath='.//li/text()' data='Python'>]
```

```
>>> sl.extract()
['C++', 'Java', 'Python']
```

SelectorList 对象还有一个 extract_first 方法,该方法返回其中第一个 Selector 对象调用 extract 方法的结果。通常,在 SelectorList 对象中只包含一个 Selector 对象时调用该方法,直接提取出 Unicode 字符串而不是列表:

```
>>> sl = selector.xpath('.//b')
>>> sl
[<Selector xpath='.//b' data='<b>Hello Python</b>'>]
>>> sl.extract()
['<b>Hello Python</b>']
>>> sl.extract_first()
'<b>Hello Python</b>'
```

有些时候,我们想使用正则表达式提取选中内容中的某部分,可以使用 re 方法(两个对象都有该方法):

```
>>> text = '''
... <ul>
...     <li>Python 学习手册 <b>价格: 99.00 元</b></li>
...     <li>Python 核心编程 <b>价格: 88.00 元</b></li>
...     <li>Python 基础教程 <b>价格: 80.00 元</b></li>
... </ul>
... '''
...
>>> selector = Selector(text=text)
>>> selector.xpath('.//li/b/text()')
[<Selector xpath='.//li/b/text()' data='价格: 99.00 元'>,
 <Selector xpath='.//li/b/text()' data='价格: 88.00 元'>,
 <Selector xpath='.//li/b/text()' data='价格: 80.00 元'>]
>>> selector.xpath('.//li/b/text()').extract()
['价格: 99.00 元', '价格: 88.00 元', '价格: 80.00 元']
>>> selector.xpath('.//li/b/text()').re('\d+\.\d+') #只提取价格的数字部分
['99.00', '88.00', '80.00']
```

SelectorList 对象的 re_first 方法同样返回其中的第一个 Selector 对象调用 re 方法的结果:

```
>>> selector.xpath('.//li/b/text()').re_first('\d+\.\d+')
'99.00'
```

3.2 Response 内置 Selector

在实际开发中,几乎不需要手动创建 Selector 对象,在第一次访问一个 Response 对象的 selector 属性时,Response 对象内部会以自身为参数自动创建 Selector 对象,并将该 Selector 对象缓存,以便下次使用。Scrapy 源码中的相关实现如下:

```python
class TextResponse(Response):

    def __init__(self, *args, **kwargs):
        ...
        self._cached_selector = None
        ...

    @property
    def selector(self):
        from scrapy.selector import Selector
        if self._cached_selector is None:
            self._cached_selector = Selector(self)
        return self._cached_selector
    ...
```

通常,我们直接使用 Response 对象内置的 Selector 对象即可:

```
>>> from scrapy.http import HtmlResponse
>>> body = '''
... <html>
...     <body>
...         <h1>Hello World</h1>
...         <h1>Hello Scrapy</h1>
...         <b>Hello python</b>
...         <ul>
...             <li>C++</li>
...             <li>Java</li>
...             <li>Python</li>
...         </ul>
...     </body>
... </html>
```

```
...'''
...
>>> response = HtmlResponse(url='http://www.example.com', body=body, encoding='utf8')
>>> response.selector
<Selector xpath=None data='<html>\n          <body>\n            <h1>He'>
```

为了方便用户使用，Response 对象还提供了 xpath 和 css 方法，它们在内部分别调用内置 Selector 对象的 xpath 和 css 方法。Scrapy 源码中的相关实现如下：

```
class TextResponse(Response):
    ...
    def xpath(self, query, **kwargs):
        return self.selector.xpath(query, **kwargs)

    def css(self, query):
        return self.selector.css(query)
    ...
```

使用这两个快捷方式可使代码更加简洁：

```
>>> response.xpath('.//h1/text()').extract()
['Hello World', 'Hello Scrapy']
>>> response.css('li::text').extract()
['C++', 'Java', 'Python']
```

3.3 XPath

XPath 即 XML 路径语言（XML Path Language），它是一种用来确定 xml 文档中某部分位置的语言。

xml 文档（html 属于 xml）是由一系列节点构成的树，例如：

```
<html>
    <body>
        <div >
            <p>Hello world<p>
            <a href="/home">Click here</a>
        </div>
    </body>
</html>
```

xml 文档的节点有多种类型，其中最常用的有以下几种：

- 根节点　整个文档树的根。
- 元素节点　html、body、div、p、a。
- 属性节点　href。
- 文本节点　Hello world、Click here。

节点间的关系有以下几种：

- 父子　body 是 html 的子节点，p 和 a 是 div 的子节点。反过来，div 是 p 和 a 的父节点。
- 兄弟　p 和 a 为兄弟节点。
- 祖先/后裔　body、div、p、a 都是 html 的后裔节点；反过来 html 是 body、div、p、a 的祖先节点。

3.3.1 基础语法

表 3-1 列出了 XPath 常用的基本语法。

表 3-1　XPath 常用的基本语法

表达式	描述
/	选中文档的根（root）
.	选中当前节点
..	选中当前节点的父节点
ELEMENT	选中子节点中所有 ELEMENT 元素节点
//ELEMENT	选中后代节点中所有 ELEMENT 元素节点
*	选中所有元素子节点
text()	选中所有文本子节点
@ATTR	选中名为 ATTR 的属性节点
@*	选中所有属性节点
[谓语]	谓语用来查找某个特定的节点或者包含某个特定值的节点

接下来，我们通过一些例子展示 XPath 的使用。

首先创建一个用于演示的 html 文档，并用其构造一个 HtmlResponse 对象：

```
>>> from scrapy.selector import Selector
>>> from scrapy.http import HtmlResponse
>>> body = '''
```

```
... <html>
...     <head>
...         <base href='http://example.com/' />
...         <title>Example website</title>
...     </head>
...     <body>
...         <div id='images'>
...             <a href='image1.html'>Name: Image 1 <br/><img src='image1.jpg' /></a>
...             <a href='image2.html'>Name: Image 2 <br/><img src='image2.jpg' /></a>
...             <a href='image3.html'>Name: Image 3 <br/><img src='image3.jpg' /></a>
...             <a href='image4.html'>Name: Image 4 <br/><img src='image4.jpg' /></a>
...             <a href='image5.html'>Name: Image 5 <br/><img src='image5.jpg' /></a>
...         </div>
...     </body>
... </html>
... '''
...
>>> response = HtmlResponse(url='http://www.example.com', body=body, encoding='utf8')
```

- /: 描述一个从根开始的绝对路径。

```
>>> response.xpath('/html')
[<Selector xpath='/html' data='<html>\n\t<head>\n\t\t<base href="http://exam'>]
>>> response.xpath('/html/head')
[<Selector xpath='/html/head' data='<head>\n\t\t<base href="http://example.com/'>]
```

- E1/E2: 选中 E1 子节点中的所有 E2。

```
# 选中 div 子节点中的所有 a
>>> response.xpath('/html/body/div/a')
[<Selector xpath='/html/body/div/a' data='<a href="image1.html">Name: My image 1 <'>,
 <Selector xpath='/html/body/div/a' data='<a href="image2.html">Name: My image 2 <'>,
 <Selector xpath='/html/body/div/a' data='<a href="image3.html">Name: My image 3 <'>,
 <Selector xpath='/html/body/div/a' data='<a href="image4.html">Name: My image 4 <'>,
 <Selector xpath='/html/body/div/a' data='<a href="image5.html">Name: My image 5 <'>]
```

- //E: 选中文档中的所有 E，无论在什么位置。

```
# 选中文档中的所有 a
>>> response.xpath('//a')
[<Selector xpath='//a' data='<a href="image1.html">Name: My image 1 <'>,
```

```
  <Selector xpath='//a' data='<a href="image2.html">Name: My image 2 <'>,
  <Selector xpath='//a' data='<a href="image3.html">Name: My image 3 <'>,
  <Selector xpath='//a' data='<a href="image4.html">Name: My image 4 <'>,
  <Selector xpath='//a' data='<a href="image5.html">Name: My image 5 <'>]
```

- E1//E2：选中 E1 后代节点中的所有 E2，无论在后代中的什么位置。

```
# 选中 body 后代中的所有 img
>>> response.xpath('/html/body//img')
[<Selector xpath='/html/body//img' data='<img src="image1.jpg">'>,
 <Selector xpath='/html/body//img' data='<img src="image2.jpg">'>,
 <Selector xpath='/html/body//img' data='<img src="image3.jpg">'>,
 <Selector xpath='/html/body//img' data='<img src="image4.jpg">'>,
 <Selector xpath='/html/body//img' data='<img src="image5.jpg">'>]
```

- E/text()：选中 E 的文本子节点。

```
# 选中所有 a 的文本
>>> sel = response.xpath('//a/text()')
>>> sel
[<Selector xpath='//a/text()' data='Name: My image 1 '>,
 <Selector xpath='//a/text()' data='Name: My image 2 '>,
 <Selector xpath='//a/text()' data='Name: My image 3 '>,
 <Selector xpath='//a/text()' data='Name: My image 4 '>,
 <Selector xpath='//a/text()' data='Name: My image 5 '>]
>>> sel.extract()
['Name: My image 1 ',
 'Name: My image 2 ',
 'Name: My image 3 ',
 'Name: My image 4 ',
 'Name: My image 5 ']
```

- E/*：选中 E 的所有元素子节点。

```
# 选中 html 的所有元素子节点
>>> response.xpath('/html/*')
[<Selector xpath='/html/*' data='<head>\n\t\t<base href="http://example.com/">,
 <Selector xpath='/html/*' data='<body>\n\t\t<div id="images">\n\t\t<a href="i'>]

# 选中 div 的所有后代元素节点
>>> response.xpath('/html/body/div//*')
```

```
[<Selector xpath='/html/body/div//*' data='<a href="image1.html">Name: My image 1 <'>,
 <Selector xpath='/html/body/div//*' data='<br>'>,
 <Selector xpath='/html/body/div//*' data='<img src="image1.jpg">'>,
 <Selector xpath='/html/body/div//*' data='<a href="image2.html">Name: My image 2 <'>,
 <Selector xpath='/html/body/div//*' data='<br>'>,
 <Selector xpath='/html/body/div//*' data='<img src="image2.jpg">'>,
 <Selector xpath='/html/body/div//*' data='<a href="image3.html">Name: My image 3 <'>,
 <Selector xpath='/html/body/div//*' data='<br>'>,
 <Selector xpath='/html/body/div//*' data='<img src="image3.jpg">'>,
 <Selector xpath='/html/body/div//*' data='<a href="image4.html">Name: My image 4 <'>,
 <Selector xpath='/html/body/div//*' data='<br>'>,
 <Selector xpath='/html/body/div//*' data='<img src="image4.jpg">'>,
 <Selector xpath='/html/body/div//*' data='<a href="image5.html">Name: My image 5 <'>,
 <Selector xpath='/html/body/div//*' data='<br>'>,
 <Selector xpath='/html/body/div//*' data='<img src="image5.jpg">'>]
```

- */E：选中孙节点中的所有 E。

```
# 选中 div 孙节点中的所有 img
>>> response.xpath('//div/*/img')
[<Selector xpath='//div/*/img' data='<img src="image1.jpg">'>,
 <Selector xpath='//div/*/img' data='<img src="image2.jpg">'>,
 <Selector xpath='//div/*/img' data='<img src="image3.jpg">'>,
 <Selector xpath='//div/*/img' data='<img src="image4.jpg">'>,
 <Selector xpath='//div/*/img' data='<img src="image5.jpg">'>]
```

- E/@ATTR：选中 E 的 ATTR 属性。

```
# 选中所有 img 的 src 属性
>>> response.xpath('//img/@src')
[<Selector xpath='//img/@src' data='image1.jpg'>,
 <Selector xpath='//img/@src' data='image2.jpg'>,
 <Selector xpath='//img/@src' data='image3.jpg'>,
 <Selector xpath='//img/@src' data='image4.jpg'>,
 <Selector xpath='//img/@src' data='image5.jpg'>]
```

- //@ATTR：选中文档中所有 ATTR 属性。

```
# 选中所有的 href 属性
>>> response.xpath('//@href')
[<Selector xpath='//@href' data='http://example.com/'>,
```

```
  <Selector xpath='//@href' data='image1.html'>,
  <Selector xpath='//@href' data='image2.html'>,
  <Selector xpath='//@href' data='image3.html'>,
  <Selector xpath='//@href' data='image4.html'>,
  <Selector xpath='//@href' data='image5.html'>]
```

- E/@*：选中 E 的所有属性。

```
# 获取第一个 a 下 img 的所有属性（这里只有 src 一个属性）
>>> response.xpath('//a[1]/img/@*')
[<Selector xpath='//a[1]/img/@*' data='image1.jpg'>]
```

- .：选中当前节点，用来描述相对路径。

```
# 获取第 1 个 a 的选择器对象
>>> sel = response.xpath('//a')[0]
>>> sel
<Selector xpath='//a' data='<a href="image1.html">Name: My image 1 <'>

# 假设我们想选中当前这个 a 后代中的所有 img，下面的做法是错误的，
# 会找到文档中所有的 img
# 因为//img 是绝对路径，会从文档的根开始搜索，而不是从当前的 a 开始
>>> sel.xpath('//img')
[<Selector xpath='//img' data='<img src="image1.jpg">'>,
 <Selector xpath='//img' data='<img src="image2.jpg">'>,
 <Selector xpath='//img' data='<img src="image3.jpg">'>,
 <Selector xpath='//img' data='<img src="image4.jpg">'>,
 <Selector xpath='//img' data='<img src="image5.jpg">'>]
# 需要使用.//img 来描述当前节点后代中的所有 img
>>> sel.xpath('.//img')
[<Selector xpath='.//img' data='<img src="image1.jpg">'>]
```

- ..：选中当前节点的父节点，用来描述相对路径。

```
# 选中所有 img 的父节点
>>> response.xpath('//img/..')
[<Selector xpath='//img/..' data='<a href="image1.html">Name: My image 1 <'>,
 <Selector xpath='//img/..' data='<a href="image2.html">Name: My image 2 <'>,
 <Selector xpath='//img/..' data='<a href="image3.html">Name: My image 3 <'>,
 <Selector xpath='//img/..' data='<a href="image4.html">Name: My image 4 <'>,
 <Selector xpath='//img/..' data='<a href="image5.html">Name: My image 5 <'>]
```

- node[谓语]: 谓语用来查找某个特定的节点或者包含某个特定值的节点。

```
# 选中所有 a 中的第 3 个
>>> response.xpath('//a[3]')
[<Selector xpath='//a[3]' data='<a href="image3.html">Name: My image 3 <>]

# 使用 last 函数，选中最后 1 个
>>> response.xpath('//a[last()]')
[<Selector xpath='//a[last()]' data='<a href="image5.html">Name: My image 5 <>]

# 使用 position 函数，选中前 3 个
>>> response.xpath('//a[position()<=3]')
[<Selector xpath='//a[position()<=3]' data='<a href="image1.html">Name: My image 1 <>,
 <Selector xpath='//a[position()<=3]' data='<a href="image2.html">Name: My image 2 <>,
 <Selector xpath='//a[position()<=3]' data='<a href="image3.html">Name: My image 3 <>]

# 选中所有含有 id 属性的 div
>>> response.xpath('//div[@id]')
[<Selector xpath='//div[@id]' data='<div id="images">\n\t\t<a href="image1.htm'>]

# 选中所有含有 id 属性且值为"images"的 div
>>> response.xpath('//div[@id="images"]')
[<Selector xpath='//div[@id="images"]' data='<div id="images">\n\t\t<a href="image1.htm'>]
```

3.3.2 常用函数

XPath 还提供许多函数，如数字、字符串、时间、日期、统计等。在上面的例子中，我们已经使用了函数 position()、last()。由于篇幅有限，下面仅介绍两个十分常用的字符串函数。

- string(arg): 返回参数的字符串值。

```
>>> from scrapy.selector import Selector
>>> text='<a href="#">Click here to go to the <strong>Next Page</strong></a>'
>>> sel = Selector(text=text)
>>> sel
<Selector xpath=None data='<html><body><a href="#">Click here to go'>
# 以下做法和 sel.xpath('/html/body/a/strong/text()')得到相同结果
>>> sel.xpath('string(/html/body/a/strong)').extract()
```

```
['Next Page']
# 如果想得到 a 中的整个字符串'Click here to go to the Next Page',
# 使用 text()就不行了，因为 Click here to go to the 和 Next Page 在不同元素下
# 以下做法将得到两个子串
>>> sel.xpath('/html/body/a//text()').extract()
['Click here to go to the ', 'Next Page']
# 这种情况下可以使用 string()函数
>>> sel.xpath('string(/html/body/a)').extract()
['Click here to go to the Next Page']
```

- contains(str1, str2)：判断 str1 中是否包含 str2，返回布尔值。

```
>>> text = '''
... <div>
...     <p class="small info">hello world</p>
...     <p class="normal info">hello scrapy</p>
... </div>
... '''
>>> sel = Selector(text=text)
>>> sel.xpath('//p[contains(@class, "small")]')  # 选择 class 属性中包含"small"的 p 元素
[<Selector xpath='//p[contains(@class, "small")]' data='<p class="small info">hello world</p>'>]
>>> sel.xpath('//p[contains(@class, "info")]')   # 选择 class 属性中包含"info"的 p 元素
[<Selector xpath='//p[contains(@class, "info")]' data='<p class="small info">hello world</p>'>,
 <Selector xpath='//p[contains(@class, "info")]' data='<p class="normal info">hello scrapy</p>'>]
```

关于 XPath 的使用先介绍到这里，更多详细内容可以参看 XPath 文档：https://www.w3.org/TR/xpath/。

3.4 CSS 选择器

CSS 即层叠样式表，其选择器是一种用来确定 HTML 文档中某部分位置的语言。

CSS 选择器的语法比 XPath 更简单一些，但功能不如 XPath 强大。实际上，当我们调用 Selector 对象的 CSS 方法时，在其内部会使用 Python 库 cssselect 将 CSS 选择器表达式翻译成 XPath 表达式，然后调用 Selector 对象的 XPATH 方法。

表 3-2 列出了 CSS 选择器的一些基本语法。

表 3-2 CSS 选择器

表 达 式	描 述	例 子
*	选中所有元素	*
E	选中 E 元素	p
E1,E2	选中 E1 和 E2 元素	div,pre
E1 E2	选中 E1 后代元素中的 E2 元素	div p
E1>E2	选中 E1 子元素中的 E2 元素	div>p
E1+E2	选中 E1 兄弟元素中的 E2 元素	p+strong
.CLASS	选中 CLASS 属性包含 CLASS 的元素	.info
#ID	选中 id 属性为 ID 的元素	#main
[ATTR]	选中包含 ATTR 属性的元素	[href]
[ATTR=VALUE]	选中包含 ATTR 属性且值为 VALUE 的元素	[method=post]
[ATTR~=VALUE]	选中包含 ATTR 属性且值包含 VALUE 的元素	[class~=clearfix]
E:nth-child(n) E:nth-last-child(n)	选中 E 元素，且该元素必须是其父元素的（倒数）第 n 个子元素	a:nth-child(1) a:nth-last-child(2)
E:first-child E:last-child	选中 E 元素，且该元素必须是其父元素的（倒数）第一个子元素	a:first-child a:last-child
E:empty	选中没有子元素的 E 元素	div:empty
E::text	选中 E 元素的文本节点（Text Node）	p::text

和学习 XPath 一样，通过一些例子展示 CSS 选择器的使用。

先创建一个 HTML 文档并构造一个 HtmlResponse 对象：

```
>>> from scrapy.selector import Selector
>>> from scrapy.http import HtmlResponse
>>> body = '''
... <html>
...     <head>
...         <base href='http://example.com/' />
...         <title>Example website</title>
...     </head>
...     <body>
...         <div id='images-1' style="width: 1230px;">
...             <a href='image1.html'>Name: Image 1 <br/><img src='image1.jpg' /></a>
...             <a href='image2.html'>Name: Image 2 <br/><img src='image2.jpg' /></a>
...             <a href='image3.html'>Name: Image 3 <br/><img src='image3.jpg' /></a>
```

```
...                    </div>
...
...            <div id='images-2' class='small'>
...                    <a href='image4.html'>Name: Image 4 <br/><img src='image4.jpg' /></a>
...                    <a href='image5.html'>Name: Image 5 <br/><img src='image5.jpg' /></a>
...            </div>
...        </body>
... </html>
... '''
...
>>> response = HtmlResponse(url='http://www.example.com', body=body, encoding='utf8')
```

- E：选中 E 元素。

```
# 选中所有的 img
>>> response.css('img')
[<Selector xpath='descendant-or-self::img' data='<img src="image1.jpg">'>,
 <Selector xpath='descendant-or-self::img' data='<img src="image2.jpg">'>,
 <Selector xpath='descendant-or-self::img' data='<img src="image3.jpg">'>,
 <Selector xpath='descendant-or-self::img' data='<img src="image4.jpg">'>,
 <Selector xpath='descendant-or-self::img' data='<img src="image5.jpg">'>]
```

- E1,E2：选中 E1 和 E2 元素。

```
# 选中所有 base 和 title
>>> response.css('base,title')
[<Selector xpath='descendant-or-self::base | descendant-or-self::title' data='<base href="http://example.com/">'>,
 <Selector xpath='descendant-or-self::base | descendant-or-self::title' data='<title>Example website</title>'>]
```

- E1 E2：选中 E1 后代元素中的 E2 元素。

```
# div 后代中的 img
>>> response.css('div img')
[<Selector xpath='descendant-or-self::div/descendant-or-self::*/img' data='<img src="image1.jpg">'>,
 <Selector xpath='descendant-or-self::div/descendant-or-self::*/img' data='<img src="image2.jpg">'>,
 <Selector xpath='descendant-or-self::div/descendant-or-self::*/img' data='<img src="image3.jpg">'>,
 <Selector xpath='descendant-or-self::div/descendant-or-self::*/img' data='<img src="image4.jpg">'>,
 <Selector xpath='descendant-or-self::div/descendant-or-self::*/img' data='<img src="image5.jpg">'>]
```

- E1>E2：选中 E1 子元素中的 E2 元素。

```
# body 子元素中的 div
>>> response.css('body>div')
[<Selector xpath='descendant-or-self::body/div' data='<div id="images-1" style="width: 1230px;'>,
 <Selector xpath='descendant-or-self::body/div' data='<div id="images-2" class="small">\n         '>]
```

- [ATTR]：选中包含 ATTR 属性的元素。

```
# 选中包含 style 属性的元素
>>> response.css('[style]')
[<Selector xpath='descendant-or-self::*[@style]' data='<div id="images-1" style="width: 1230px;'>]
```

- [ATTR=VALUE]：选中包含 ATTR 属性且值为 VALUE 的元素。

```
# 选中属性 id 值为 images-1 的元素
>>> response.css('[id=images-1]')
[<Selector xpath="descendant-or-self::*[@id = 'images-1']" data='<div id="images-1" style="width: 1230px;'>]
```

- E:nth-child(n)：选中 E 元素，且该元素必须是其父元素的第 n 个子元素。

```
# 选中每个 div 的第一个 a
>>> response.css('div>a:nth-child(1)')
[<Selector xpath="descendant-or-self::div/*[name() = 'a' and (position() = 1)]" data='<a href="image1.html">Name: Image 1 <br>'>,
 <Selector xpath="descendant-or-self::div/*[name() = 'a' and (position() = 1)]" data='<a href="image4.html">Name: Image 4 <br>'>]

# 选中第二个 div 的第一个 a
>>> response.css('div:nth-child(2)>a:nth-child(1)')
[<Selector xpath="descendant-or-self::*/*[name() = 'div' and (position() = 2)]/*[name() = 'a' and (position() = 1)]" data='<a href="image4.html">Name: Image 4 <br>'>]
```

- E:first-child：选中 E 元素，该元素必须是其父元素的第一个子元素。
- E:last-child：选中 E 元素，该元素必须是其父元素的倒数第一个子元素。

```
# 选中第一个 div 的最后一个 a
>>> response.css('div:first-child>a:last-child')
[<Selector xpath="descendant-or-self::*/*[name() = 'div' and (position() = 1)]/*[name() = 'a' and (position() = last())]" data='<a href="image3.html">Name: Image 3 <br>'>]
```

- E::text: 选中 E 元素的文本节点。

```
# 选中所有 a 的文本
>>> sel = response.css('a::text')
>>> sel
[<Selector xpath='descendant-or-self::a/text()' data='Name: Image 1 '>,
 <Selector xpath='descendant-or-self::a/text()' data='Name: Image 2 '>,
 <Selector xpath='descendant-or-self::a/text()' data='Name: Image 3 '>,
 <Selector xpath='descendant-or-self::a/text()' data='Name: Image 4 '>,
 <Selector xpath='descendant-or-self::a/text()' data='Name: Image 5 '>]
>>> sel.extract()
['Name: Image 1 ',
 'Name: Image 2 ',
 'Name: Image 3 ',
 'Name: Image 4 ',
 'Name: Image 5 ']
```

关于 CSS 选择器的使用先介绍到这里，更多详细内容可以参看 CSS 选择器文档：https://www.w3.org/TR/css3-selectors/。

3.5 本章小结

本章学习了从页面中提取数据的相关内容，首先带大家了解了 Scrapy 中的 Selector 对象，然后学习如何使用 Selector 对象在页面中选中并提取数据，最后通过一系列例子讲解了 XPath 和 CSS 选择器的用法。

第 4 章

使用 Item 封装数据

在第 3 章中，我们学习了从页面中提取数据的方法，本章来学习如何封装爬取到的数据。以爬取某图书网站的书籍信息为例，对于网站中的每一本书可以提取出书名、价格、作者、出版社、出版时间等多个信息字段。应该用怎样的数据结构来维护这些零散的信息字段呢？最容易想到是使用 Python 字典（dict）。

回顾第 1 章 example 项目中 BooksSpider 的代码：

```
class BooksSpider(scrapy.Spider):
    ...
    def parse(self, response):
        for sel in response.css('article.product_pod'):
            name = sel.xpath('./h3/a/@title').extract_first()
            price = sel.css('p.price_color::text').extract_first()
            yield {
                'name': name,
                'price': price,
            }
    ...
```

在该案例中，我们便使用了 Python 字典存储一本书的信息，但字典可能有以下缺点：

(1) 无法一目了然地了解数据中包含哪些字段，影响代码可读性。
(2) 缺乏对字段名字的检测，容易因程序员的笔误而出错。
(3) 不便于携带元数据（传递给其他组件的信息）。

为解决上述问题，在 Scrapy 中可以使用自定义的 Item 类封装爬取到的数据。

4.1 Item 和 Field

Scrapy 提供了以下两个类，用户可以使用它们自定义数据类（如书籍信息），封装爬取到的数据：

- Item 基类
  自定义数据类（如 BookItem）的基类。
- Field 类
  用来描述自定义数据类包含哪些字段（如 name、price 等）。

自定义一个数据类，只需继承 Item，并创建一系列 Field 对象的类属性（类似于在 Django 中自定义 Model）即可。以定义书籍信息 BookItem 为例，它包含两个字段，分别为书的名字 name 和书的价格 price，代码如下：

```
>>> from scrapy import Item, Field
>>> class BookItem(Item):
...     name = Field()
...     price = Field()
```

Item 支持字典接口，因此 BookItem 在使用上和 Python 字典类似，可按以下方式创建 BookItem 对象：

```
>>> book1 = BookItem(name='Needful Things', price=45.0)
>>> book1
{'name': 'Needful Things', 'price': 45.0}
>>> book2 = BookItem()
>>> book2
{}
>>> book2['name'] = 'Life of Pi'
```

```
>>> book2['price'] = 32.5
{'name': 'Life of Pi', 'price': 32.5}
```

对字段进行赋值时，BookItem 内部会对字段名进行检测，如果赋值一个没有定义的字段，就会抛出异常（防止因用户粗心而导致错误）：

```
>>> book = BookItem()
>>> book['name'] = 'Memoirs of a Geisha'
>>> book['prize'] = 43.0      # 粗心，把 price 拼写成了 prize.
Traceback (most recent call last):
    ...
KeyError: 'BookItem does not support field: prize'
```

访问 BookItem 对象中的字段与访问字典类似，示例如下：

```
>>> book = BookItem(name='Needful Things', price=45.0)
>>> book['name']
'Needful Things'
>>> book.get('price', 60.0)
45.0
>>> list(book.items())
[('price', 45.0), ('name', 'Needful Things')]
```

接下来，我们改写第 1 章 example 项目中的代码，使用 Item 和 Field 定义 BookItem 类，用其封装爬取到的书籍信息项目目录下的 items.py 文件供用户实现各种自定义的数据类，在 items.py 中实现 BookItem，代码如下：

```
from scrapy import Item, Field

class BookItem(Item):
    name = Field()
    price = Field()
```

修改之前的 BooksSpider，使用 BookItem 替代 Python 字典，代码如下：

```
from ..items import BookItem

class BooksSpider(scrapy.Spider):
    ...
    def parse(self, response):
        for sel in response.css('article.product_pod'):
            book = BookItem()
```

```
        book['name'] = sel.xpath('./h3/a/@title').extract_first()
        book['price'] = sel.css('p.price_color::text').extract_first()
        yield book
    ...
```

4.2 拓展 Item 子类

有些时候，我们可能要根据需求对已有的自定义数据类（Item 子类）进行拓展。例如，example 项目中又添加了一个新的 Spider，它负责在另外的图书网站爬取国外书籍（中文翻译版）的信息，此类书籍的信息比之前多了一个译者字段，此时可以继承 BookItem 定义一个 ForeignBookItem 类，在其中添加一个译者字段，代码如下：

```
>>> class ForeignBookItem(BookItem):
...     translator = Field()
...
>>> book = ForeignBookItem()
>>> book['name'] = '巴黎圣母院'
>>> book['price'] = 20.0
>>> book['translator'] = '陈敬容'
```

4.3 Field 元数据

在第 2 章中曾讲到，一项数据由 Spider 提交给 Scrapy 引擎后，可能会被递送给其他组件（Item Pipeline、Exporter）处理。假设想传递额外信息给处理数据的某个组件（例如，告诉该组件应以怎样的方式处理数据），此时可以使用 Field 的元数据。请看下面的例子：

```
class ExampleItem(Item):
    x = Field(a='hello', b=[1, 2, 3])        # x 有两个元数据，a 是个字符串，b 是个列表
    y = Field(a=lambda x: x ** 2)            # y 有一个元数据，a 是个函数
```

访问一个 ExampleItem 对象的 fields 属性，将得到一个包含所有 Field 对象的字典：

```
>>> e = ExampleItem(x=100, y=200)
>>> e.fields
{'x': {'a': 'hello', 'b': [1, 2, 3]},
```

```
'y': {'a': <function __main__.ExampleItem.<lambda>>}}
>>> type(e.fields['x'])
scrapy.item.Field
>>> type(e.fields['y'])
scrapy.item.Field
```

实际上，Field 是 Python 字典的子类，可以通过键获取 Field 对象中的元数据：

```
>>> issubclass(Field, dict)
True
>>> field_x = e.fields['x']        # 注意，不要混淆 e.fields['x']和 e['x']
>>> field_x
{'a': 'hello', 'b': [1, 2, 3]}
>>> field_x['a']
'hello'
>>> field_y = e.fields['y']
>>> field_y
{'a': <function __main__.ExampleItem.<lambda>>}
>>> field_y.get('a', lambda x: x)
<function __main__.ExampleItem.<lambda>>
```

接下来，看一个应用 Field 元数据的实际例子。假设我们要把爬取到的书籍信息写入 csv 文件，那每一项数据最终由 Scrapy 提供的 CsvItemExporter 写入文件（数据导出在第 7 章详细讲解），在爬取过程中提取到的信息并不总是一个字符串，有时可能是一个字符串列表，例如：

```
>>> book['authors'] = ['李雷', '韩梅梅', '吉姆']
```

但在写入 csv 文件时，需要将列表内所有字符串串行化成一个字符串，串行化的方式有很多种，例如：

```
1. '李雷|韩梅梅|吉姆'            # '|'.join(book['authors'])
2. '李雷;韩梅梅;吉姆'            # ';'.join(book['authors'])
3. "['李雷', '韩梅梅', '吉姆']"   # str(book['authors'])
```

我们可以通过 authors 字段的元数据告诉 CsvItemExporter 如何对 authors 字段串行化：

```
class BookItem(Item):
    ...
    authors = Field(serializer=lambda x: '|'.join(x))
    ...
```

其中，元数据的键 serializer 是 CsvItemExporter 规定好的，它会用该键获取元数据，即一个串行化函数对象，并使用这个串行化函数将 authors 字段串行化成一个字符串。以下是 Scrapy 源码中的相关实现：

```python
# exports.py

class BaseItemExporter(object):
    ...

    def _get_serialized_fields(self, item, default_value=None, include_empty=None):
        ...
        for field_name in field_iter:
            if field_name in item:
                field = {} if isinstance(item, dict) else item.fields[field_name]
                value = self.serialize_field(field, field_name, item[field_name])
            else:
                value = default_value

            yield field_name, value
    ...

class CsvItemExporter(BaseItemExporter):
    ...

    def export_item(self, item):
        ...
        fields = self._get_serialized_fields(item, default_value='',
                                              include_empty=True)
        values = list(self._build_row(x for _, x in fields))
        self.csv_writer.writerow(values)
    ...

    def serialize_field(self, field, name, value):
        serializer = field.get('serializer', self._join_if_needed)
        return serializer(value)
    ...
```

解释上述代码如下：

- 爬取到的每一项数据由 export_item 方法导出到文件，写入文件之前，先调用 _get_serialized_fields 方法（在基类中实现）获得数据中每个字段串行化的结果。
- 在 _get_serialized_fields 方法中调用 serialize_field 方法，获取其中一个字段串行化的结果。
- 在 serialize_field 方法中获取字段的元数据 serializer，得到串行化函数（如果不存在，就使用默认的 _join_if_needed 函数），最终调用该函数对字段串行化，并将结果返回。

在实际应用中，我们可以仿照上面的例子灵活使用 Field 元数据。

4.4 本章小结

本章介绍了在 Scrapy 中如何封装爬取到的数据，先了解了 Item 基类以及用来定义字段的 Field 类，然后展示了一个使用它们封装数据的例子。最后，还介绍了使用 Field 元数据给其他组件传递信息的方法。

第 5 章

使用 Item Pipeline 处理数据

在之前的章节中，我们学习了提取数据以及封装数据的方法，这一章来学习如何对爬取到的数据进行处理。在 Scrapy 中，Item Pipeline 是处理数据的组件，一个 Item Pipeline 就是一个包含特定接口的类，通常只负责一种功能的数据处理，在一个项目中可以同时启用多个 Item Pipeline，它们按指定次序级联起来，形成一条数据处理流水线。

以下是 Item Pipeline 的几种典型应用：

- 清洗数据。
- 验证数据的有效性。
- 过滤掉重复的数据。
- 将数据存入数据库。

5.1 Item Pipeline

通过一个例子讲解 Item Pipeline 的使用，在第 1 章的 example 项目中，我们爬取到的书籍价格是以英镑为单位的：

```
$ scrapy crawl books -o books.csv
...
$ head -5 books.csv        # 查看文件开头的 5 行
name,price
A Light in the Attic,£51.77
Tipping the Velvet,£53.74
Soumission,£50.10
Sharp Objects,£47.82
```

如果我们期望爬取到的书价是人民币价格，就需要用英镑价格乘以汇率计算出人民币价格（处理数据），此时可以实现一个价格转换的 Item Pipeline 来完成这个工作。接下来在 example 项目中实现它。

5.1.1 实现 Item Pipeline

在创建一个 Scrapy 项目时，会自动生成一个 pipelines.py 文件，它用来放置用户自定义的 Item Pipeline，在 example 项目的 pipelines.py 中实现 PriceConverterPipeline，代码如下：

```python
class PriceConverterPipeline(object):

    # 英镑兑换人民币汇率
    exchange_rate = 8.5309

    def process_item(self, item, spider):
        # 提取 item 的 price 字段（如£53.74）
        # 去掉前面英镑符号£，转换为 float 类型，乘以汇率
        price = float(item['price'][1:]) * self.exchange_rate

        # 保留 2 位小数，赋值回 item 的 price 字段
        item['price'] = '¥%.2f' % price

        return item
```

对上述代码解释如下：

- 一个 Item Pipeline 不需要继承特定基类，只需要实现某些特定方法，例如 process_item、open_spider、close_spider。

- 一个 Item Pipeline 必须实现一个 process_item(item, spider)方法，该方法用来处理每一项由 Spider 爬取到的数据，其中的两个参数：
  - Item 爬取到的一项数据（Item 或字典）。
  - Spider 爬取此项数据的 Spider 对象。

上述代码中的 process_item 方法实现非常简单，将书籍的英镑价格转换为浮点数，乘以汇率并保留 2 位小数，然后赋值回 item 的 price 字段，最后返回被处理过的 item。

可以看出，process_item 方法是 Item Pipeline 的核心，对该方法还需再做两点补充说明：

- 如果 process_item 在处理某项 item 时返回了一项数据（Item 或字典），返回的数据会递送给下一级 Item Pipeline（如果有）继续处理。
- 如果 process_item 在处理某项 item 时抛出（raise）一个 DropItem 异常（scrapy.exceptions.DropItem），该项 item 便会被抛弃，不再递送给后面的 Item Pipeline 继续处理，也不会导出到文件。通常，我们在检测到无效数据或想要过滤数据时，抛出 DropItem 异常。

除了必须实现的 process_item 方法外，还有 3 个比较常用的方法，可根据需求选择实现：

- open_spider(self, spider)
  Spider 打开时（处理数据前）回调该方法，通常该方法用于在开始处理数据之前完成某些初始化工作，如连接数据库。
- close_spider(self, spider)
  Spider 关闭时（处理数据后）回调该方法，通常该方法用于在处理完所有数据之后完成某些清理工作，如关闭数据库。
- from_crawler(cls, crawler)
  创建 Item Pipeline 对象时回调该类方法。通常，在该方法中通过 crawler.settings 读取配置，根据配置创建 Item Pipeline 对象。

在后面的例子中，我们展示了以上方法的应用场景。

5.1.2 启用 Item Pipeline

在 Scrapy 中，Item Pipeline 是可选的组件，想要启用某个（或某些）Item Pipeline，需要在配置文件 settings.py 中进行配置：

```
ITEM_PIPELINES = {
    'example.pipelines.PriceConverterPipeline': 300,
}
```

ITEM_PIPELINES 是一个字典，我们把想要启用的 Item Pipeline 添加到这个字典中，其中每一项的键是每一个 Item Pipeline 类的导入路径，值是一个 0~1000 的数字，同时启用多个 Item Pipeline 时，Scrapy 根据这些数值决定各 Item Pipeline 处理数据的先后次序，数值小的在前。

启用 PriceConverterPipeline 后，重新运行爬虫，并观察结果：

```
$ scrapy crawl books -o books.csv
...
$ head -5 books.csv  # 查看文件开头的 5 行
name,price
A Light in the Attic,¥441.64
Tipping the Velvet,¥458.45
Soumission,¥427.40
Sharp Objects,¥407.95
```

使用 PriceConverterPipeline 对数据进行处理后，books.csv 中的书价转换成了人民币价格。

5.2 更多例子

我们通过一个例子学习了如何使用 Item Pipeline 对数据进行处理，下面再看两个实际例子。

5.2.1 过滤重复数据

为了确保爬取到的书籍信息中没有重复项，可以实现一个去重 Item Pipeline。这里，我们就以书名作为主键（实际应以 ISBN 编号为主键，但是仅爬取了书名和价格）进行去重，实现 DuplicatesPipeline 代码如下：

```
from scrapy.exceptions import DropItem

class DuplicatesPipeline(object):

    def __init__(self):
```

```python
        self.book_set = set()

    def process_item(self, item, spider):
        name = item['name']
        if name in self.book_set:
            raise DropItem("Duplicate book found: %s" % item)

        self.book_set.add(name)
        return item
```

对上述代码解释如下：

- 增加构造器方法，在其中初始化用于对书名去重的集合。
- 在 process_item 方法中，先取出 item 的 name 字段，检查书名是否已在集合 book_set 中，如果存在，就是重复数据，抛出 DropItem 异常，将 item 抛弃；否则，将 item 的 name 字段存入集合，返回 item。

接下来测试 DuplicatesPipeline。首先在不启用 DuplicatesPipeline 的情况下，运行爬虫，查看结果：

```
$ scrapy crawl books -o book1.csv
...
$ cat -n book1.csv
     1  price,name
     2  ¥441.64,A Light in the Attic
     3  ¥458.45,Tipping the Velvet
     4  ¥427.40,Soumission
     5  ¥407.95,Sharp Objects
     6  ¥462.63,Sapiens: A Brief History of Humankind
     7  ¥193.22,The Requiem Red
     8  ¥284.42,The Dirty Little Secrets of Getting Your Dream Job
     ...
   993  ¥317.86,Bounty (Colorado Mountain #7)
   994  ¥173.18,Blood Defense (Samantha Brinkman #1)
   995  ¥295.60,"Bleach, Vol. 1: Strawberry and the Soul Reapers (Bleach #1)"
   996  ¥370.07,Beyond Good and Evil
   997  ¥473.72,Alice in Wonderland (Alice's Adventures in Wonderland #1)
   998  ¥486.77,"Ajin: Demi-Human, Volume 1 (Ajin: Demi-Human #1)"
   999  ¥144.77,A Spy's Devotion (The Regency Spies of London #1)
```

```
     1000   ¥460.50,1st to Die (Women's Murder Club #1)
     1001   ¥222.49,"1,000 Places to See Before You Die"
```

此时有 1000 本书。

然后在配置文件 settings.py 中启用 DuplicatesPipeline：

```
ITEM_PIPELINES = {
    'example.pipelines.PriceConverterPipeline': 300,
    'example.pipelines.DuplicatesPipeline': 350,
}
```

运行爬虫，对比结果：

```
$ scrapy crawl books -o book2.csv
...
$ cat -n book2.csv
     1   name,price
     2   A Light in the Attic,¥441.64
     3   Tipping the Velvet,¥458.45
     4   Soumission,¥427.40
     5   Sharp Objects,¥407.95
     6   Sapiens: A Brief History of Humankind,¥462.63
     7   The Requiem Red,¥193.22
     8   The Dirty Little Secrets of Getting Your Dream Job,¥284.42
     ...
     993  Blood Defense (Samantha Brinkman #1),¥173.18
     994  "Bleach, Vol. 1: Strawberry and the Soul Reapers (Bleach #1)",¥295.60
     995  Beyond Good and Evil,¥370.07
     996  Alice in Wonderland (Alice's Adventures in Wonderland #1),¥473.72
     997  "Ajin: Demi-Human, Volume 1 (Ajin: Demi-Human #1)",¥486.77
     998  A Spy's Devotion (The Regency Spies of London #1),¥144.77
     999  1st to Die (Women's Murder Club #1),¥460.50
     1000 "1,000 Places to See Before You Die",¥222.49
```

只有 999 本了，比之前少了 1 本，说明有两本书是同名的，翻阅爬虫的 log 信息可以找到重复项：

```
[scrapy.core.scraper] WARNING: Dropped: Duplicate book found:
{'name': 'The Star-Touched Queen', 'price': '¥275.55'}
```

5.2.2 将数据存入 MongoDB

有时,我们想把爬取到的数据存入某种数据库中,可以实现 Item Pipeline 完成此类任务。下面实现一个能将数据存入 MongoDB 数据库的 Item Pipeline,代码如下:

```python
from scrapy.item import Item
import pymongo

class MongoDBPipeline(object):

    DB_URI = 'mongodb://localhost:27017/'
    DB_NAME = 'scrapy_data'

    def open_spider(self, spider):
        self.client = pymongo.MongoClient(self.DB_URI)
        self.db = self.client[self.DB_NAME]

    def close_spider(self, spider):
        self.client.close()

    def process_item(self, item, spider):
        collection = self.db[spider.name]
        post = dict(item) if isinstance(item, Item) else item
        collection.insert_one(post)
        return item
```

对上述代码解释如下。

- 在类属性中定义两个常量:
  - DB_URI 数据库的 URI 地址。
  - DB_NAME 数据库的名字。
- 在 Spider 整个爬取过程中,数据库的连接和关闭操作只需要进行一次,应在开始处理数据之前连接数据库,并在处理完所有数据之后关闭数据库。因此实现以下两个方法(在 Spider 打开和关闭时被调用):
  - open_spider(spider)
  - close_spider(spider)

  分别在 open_spider 和 close_spider 方法中实现数据库的连接与关闭。

- 在 process_item 中实现 MongoDB 数据库的写入操作，使用 self.db 和 spider.name 获取一个集合（collection），然后将数据插入该集合，集合对象的 insert_one 方法需传入一个字典对象（不能传入 Item 对象），因此在调用前先对 item 的类型进行判断，如果 item 是 Item 对象，就将其转换为字典。

接下来测试 MongoDBPipeline，在配置文件 settings.py 中启用 MongoDBPipeline：

```
ITEM_PIPELINES = {
    'example.pipelines.PriceConverterPipeline': 300,
    'example.pipelines.MongoDBPipeline': 400,
}
```

运行爬虫，并查看数据库中的结果：

```
$ scrapy crawl books
...
$ mongo
MongoDB shell version: 2.4.9
connecting to: test
> use scrapy_data
switched to db scrapy_data
> db.books.count()
1000
> db.books.find()
{ "_id" : ObjectId("58ae39a89dcd191973cc588f"), "price" : "￥441.64", "name" : "A Light in the Attic" }
{ "_id" : ObjectId("58ae39a89dcd191973cc5890"), "price" : "￥458.45", "name" : "Tipping the Velvet" }
{ "_id" : ObjectId("58ae39a89dcd191973cc5891"), "price" : "￥427.40", "name" : "Soumission" }
{ "_id" : ObjectId("58ae39a89dcd191973cc5892"), "price" : "￥407.95", "name" : "Sharp Objects" }
{ "_id" : ObjectId("58ae39a89dcd191973cc5893"), "price" : "￥462.63", "name" : "Sapiens: A Brief History of Humankind" }
{ "_id" : ObjectId("58ae39a89dcd191973cc5894"), "price" : "￥193.22", "name" : "The Requiem Red" }
{ "_id" : ObjectId("58ae39a89dcd191973cc5895"), "price" : "￥284.42", "name" : "The Dirty Little Secrets of Getting Your Dream Job" }
{ "_id" : ObjectId("58ae39a89dcd191973cc5896"), "price" : "￥152.96", "name" : "The Coming Woman: A Novel Based on the Life of the Infamous Feminist, Victoria Woodhull" }
{ "_id" : ObjectId("58ae39a89dcd191973cc5897"), "price" : "￥192.80", "name" : "The Boys in the Boat: Nine Americans and Their Epic Quest for Gold at the 1936 Berlin Olympics" }
```

{ "_id" : ObjectId("58ae39a89dcd191973cc5898"), "price" : "¥444.89", "name" : "The Black Maria" }
{ "_id" : ObjectId("58ae39a89dcd191973cc5899"), "price" : "¥119.35", "name" : "Starving Hearts (Triangular Trade Trilogy, #1)" }
{ "_id" : ObjectId("58ae39a89dcd191973cc589a"), "price" : "¥176.25", "name" : "Shakespeare's Sonnets" }
{ "_id" : ObjectId("58ae39a89dcd191973cc589b"), "price" : "¥148.95", "name" : "Set Me Free" }
{ "_id" : ObjectId("58ae39a89dcd191973cc589c"), "price" : "¥446.08", "name" : "Scott Pilgrim's Precious Little Life (Scott Pilgrim #1)" }
{ "_id" : ObjectId("58ae39a89dcd191973cc589d"), "price" : "¥298.75", "name" : "Rip it Up and Start Again" }
{ "_id" : ObjectId("58ae39a89dcd191973cc589e"), "price" : "¥488.39", "name" : "Our Band Could Be Your Life: Scenes from the American Indie Underground, 1981-1991" }
{ "_id" : ObjectId("58ae39a89dcd191973cc589f"), "price" : "¥203.72", "name" : "Olio" }
{ "_id" : ObjectId("58ae39a89dcd191973cc58a0"), "price" : "¥320.68", "name" : "Mesaerion: The Best Science Fiction Stories 1800-1849" }
{ "_id" : ObjectId("58ae39a89dcd191973cc58a1"), "price" : "¥437.89", "name" : "Libertarianism for Beginners" }
{ "_id" : ObjectId("58ae39a89dcd191973cc58a2"), "price" : "¥385.34", "name" : "It's Only the Himalayas" }
Type "it" for more

在上述实现中，数据库的 URI 地址和数据库的名字硬编码在代码中，如果希望通过配置文件设置它们，只需稍作改动，代码如下：

```python
from scrapy.item import Item
import pymongo

class MongoDBPipeline(object):
    @classmethod
    def from_crawler(cls, crawler):
        cls.DB_URI = crawler.settings.get('MONGO_DB_URI',
                                          'mongodb://localhost:27017/')
        cls.DB_NAME = crawler.settings.get('MONGO_DB_NAME', 'scrapy_data')

        return cls()

    def open_spider(self, spider):
        self.client = pymongo.MongoClient(self.DB_URI)
        self.db = self.client[self.DB_NAME]
```

```
    def close_spider(self, spider):
        self.client.close()

    def process_item(self, item, spider):
        collection = self.db[spider.name]
        post = dict(item) if isinstance(item, Item) else item
        collection.insert_one(post)

        return item
```

对上述改动解释如下：

- 增加类方法 from_crawler(cls, crawler)，替代在类属性中定义 DB_URI 和 DB_NAME。
- 如果一个 Item Pipeline 定义了 from_crawler 方法，Scrapy 就会调用该方法来创建 Item Pipeline 对象。该方法有两个参数：
  - cls   Item Pipeline 类的对象（这里为 MongoDBPipeline 类对象）。
  - crawler   Crawler 是 Scrapy 中的一个核心对象，可以通过 crawler 的 settings 属性访问配置文件。
- 在 from_crawler 方法中，读取配置文件中的 MONGO_DB_URI 和 MONGO_DB_NAME（不存在使用默认值），赋给 cls 的属性，即 MongoDBPipeline 类属性。
- 其他代码并没有任何改变，因为这里只是改变了设置 MongoDBPipeline 类属性的方式。

现在，我们可在配置文件 settings.py 中对所要使用的数据库进行设置：

```
MONGO_DB_URI = 'mongodb://192.168.1.105:27017/'
MONGO_DB_NAME = 'liushuo_scrapy_data'
```

5.3 本章小结

本章学习了如何使用 Item Pipeline 对爬取到的数据进行处理，先以一个简单的例子讲解了 Item Pipeline 的应用场景以及具体使用，然后展示了 Item Pipeline 实际应用的两个例子。

第 6 章

使用 LinkExtractor 提取链接

在爬取一个网站时,想要爬取的数据通常分布在多个页面中,每个页面包含一部分数据以及到其他页面的链接,提取页面中数据的方法大家已经掌握,提取链接有使用 Selector 和使用 LinkExtractor 两种方法。

本章来学习如何提取页面中的链接。

1. 使用 Selector

因为链接也是页面中的数据,所以可以使用与提取数据相同的方法进行提取,在提取少量(几个)链接或提取规则比较简单时,使用 Selector 就足够了。

2. 使用 LinkExtractor

Scrapy 提供了一个专门用于提取链接的类 LinkExtractor,在提取大量链接或提取规则比较复杂时,使用 LinkExtractor 更加方便。

在第 1 章的 example 项目中使用了第一种方法提取下一页链接,回顾其中的代码片段:

```
class BooksSpider(scrapy.Spider):
    ...
```

```python
def parse(self, response):
    ...
    # 提取链接
    # 下一页的 url 在 ul.pager > li.next > a 里面
    # 例如： <li class="next"><a href="catalogue/page-2.html">next</a></li>
    next_url = response.css('ul.pager li.next a::attr(href)').extract_first()
    if next_url:
        # 如果找到下一页的 url，得到绝对路径，构造新的 Request 对象
        next_url = response.urljoin(next_url)
        yield scrapy.Request(next_url, callback=self.parse)
    ...
```

上述代码中，先使用 CSS 选择器选中包含下一页链接的 a 元素并获取其 href 属性，然后调用 response.urljoin 方法计算出绝对 url 地址，最后构造 Request 对象并提交。

第一种方法大家早已掌握，本章我们来学习如何使用 LinkExtractor 提取链接。

6.1　使用 LinkExtractor

LinkExtractor 的使用非常简单，通过一个例子进行讲解，使用 LinkExtractor 替代 Selector 完成 BooksSpider 提取链接的任务，代码如下：

```python
from scrapy.linkextractors import LinkExtractor
class BooksSpider(scrapy.Spider):
    ...
    def parse(self, response):
        ...
        # 提取链接
        # 下一页的 url 在 ul.pager > li.next > a 里面
        # 例如: <li class="next"><a href="catalogue/page-2.html">next</a></li>
        le = LinkExtractor(restrict_css='ul.pager li.next')
        links = le.extract_links(response)
        if links:
            next_url = links[0].url
            yield scrapy.Request(next_url, callback=self.parse)
```

对上述代码解释如下：

- 导入 LinkExtractor，它位于 scrapy.linkextractors 模块。
- 创建一个 LinkExtractor 对象，使用一个或多个构造器参数描述提取规则，这里传递给 restrict_css 参数一个 CSS 选择器表达式。它描述出下一页链接所在的区域（在 li.next 下）。
- 调用 LinkExtractor 对象的 extract_links 方法传入一个 Response 对象，该方法依据创建对象时所描述的提取规则，在 Response 对象所包含的页面中提取链接，最终返回一个列表，其中的每一个元素都是一个 Link 对象，即提取到的一个链接。
- 由于页面中的下一页链接只有一个，因此用 links[0] 获取 Link 对象，Link 对象的 url 属性便是链接页面的绝对 url 地址（无须再调用 response.urljoin 方法），用其构造 Request 对象并提交。

通过上面的例子，相信大家已经了解了使用 LinkExtractor 对象提取页面中链接的流程。

6.2　描述提取规则

接下来，我们来学习使用 LinkExtractor 的构造器参数描述提取规则。

为了在讲解过程中举例，首先制造一个实验环境，创建两个包含多个链接的 HTML 页面：

```html
<!-- example1.html -->
<html>
    <body>
        <div id="top">
            <p>下面是一些站内链接</p>
            <a class="internal" href="/intro/install.html">Installation guide</a>
            <a class="internal" href="/intro/tutorial.html">Tutorial</a>
            <a class="internal" href="../examples.html">Examples</a>
        </div>
        <div id="bottom">
            <p>下面是一些站外链接</p>
            <a href="http://stackoverflow.com/tags/scrapy/info">StackOverflow</a>
            <a href="https://github.com/scrapy/scrapy">Fork on Github</a>
        </div>
```

```
    </body>
</html>

<!-- example2.html -->
<html>
    <head>
        <script type='text/javascript' src='/js/app1.js'/>
        <script type='text/javascript' src='/js/app2.js'/>
    </head>
    <body>
        <a href="/home.html">主页</a>
        <a href="javascript:goToPage('/doc.html'); return false">文档</a>
        <a href="javascript:goToPage('/example.html'); return false">案例</a>
    </body>
</html>
```

使用以上两个 HTML 文本构造两个 Response 对象：

```
>>> from scrapy.http import HtmlResponse
>>> html1 = open('exmaple1.html').read()
>>> html2 = open('exmaple2.html').read()
>>> response1 = HtmlResponse(url='http://example1.com', body=html1, encoding='utf8')
>>> response2 = HtmlResponse(url='http://example2.com', body=html2, encoding='utf8')
```

现在有了实验环境，先说明一种特例情况，LinkExtractor 构造器的所有参数都有默认值，如果构造对象时不传递任何参数（使用默认值），就提取页面中所有链接。以下代码将提取页面 example1.html 中的所有链接：

```
>>> from scrapy.linkextractors import LinkExtractor
>>> le = LinkExtractor()
>>> links = le.extract_links(response1)
>>> [link.url for link in links]
['http://example1.com/intro/install.html',
 'http://example1.com/intro/tutorial.html',
 'http://example1.com/../examples.html',
 'http://stackoverflow.com/tags/scrapy/info',
 'https://github.com/scrapy/scrapy']
```

下面依次介绍 LinkExtractor 构造器的各个参数：

- allow

  接收一个正则表达式或一个正则表达式列表,提取绝对 url 与正则表达式匹配的链接,如果该参数为空(默认),就提取全部链接。

  示例 提取页面 example1.html 中路径以/intro 开始的链接:

  ```
  >>> from scrapy.linkextractors import LinkExtractor
  >>> pattern = '/intro/.+\.html$'
  >>> le = LinkExtractor(allow=pattern)
  >>> links = le.extract_links(response1)
  >>> [link.url for link in links]
  ['http://example1.com/intro/install.html',
   'http://example1.com/intro/tutorial.html']
  ```

- deny

  接收一个正则表达式或一个正则表达式列表,与 allow 相反,排除绝对 url 与正则表达式匹配的链接。

  示例 提取页面 example1.html 中所有站外链接(即排除站内链接):

  ```
  >>> from scrapy.linkextractors import LinkExtractor
  >>> from urllib.parse import urlparse
  >>> pattern = patten = '^' + urlparse(response1.url).geturl()
  >>> pattern
  '^http://example1.com'
  >>> le = LinkExtractor(deny=pattern)
  >>> links = le.extract_links(response1)
  >>> [link.url for link in links]
  ['http://stackoverflow.com/tags/scrapy/info',
   'https://github.com/scrapy/scrapy']
  ```

- allow_domains

  接收一个域名或一个域名列表,提取到指定域的链接。

  示例 提取页面 example1.html 中所有到 github.com 和 stackoverflow.com 这两个域的链接:

  ```
  >>> from scrapy.linkextractors import LinkExtractor
  >>> domains = ['github.com', 'stackoverflow.com']
  >>> le = LinkExtractor(allow_domains=domains)
  >>> links = le.extract_links(response1)
  >>> [link.url for link in links]
  ```

```
['http://stackoverflow.com/tags/scrapy/info',
 'https://github.com/scrapy/scrapy']
```

- deny_domains
  接收一个域名或一个域名列表，与 allow_domains 相反，排除到指定域的链接。
  示例 提取页面 example1.html 中除了到 github.com 域以外的链接：

```
>>> from scrapy.linkextractors import LinkExtractor
>>> le = LinkExtractor(deny_domains='github.com')
>>> links = le.extract_links(response1)
>>> [link.url for link in links]
['http://example1.com/intro/install.html',
 'http://example1.com/intro/tutorial.html',
 'http://example1.com/../examples.html',
 'http://stackoverflow.com/tags/scrapy/info']
```

- restrict_xpaths
  接收一个 XPath 表达式或一个 XPath 表达式列表，提取 XPath 表达式选中区域下的链接。
  示例 提取页面 example1.html 中<div id="top">元素下的链接：

```
>>> from scrapy.linkextractors import LinkExtractor
>>> le = LinkExtractor(restrict_xpaths='//div[@id="top"]')
>>> links = le.extract_links(response1)
>>> [link.url for link in links]
['http://example1.com/intro/install.html',
 'http://example1.com/intro/tutorial.html',
 'http://example1.com/../examples.html']
```

- restrict_css
  接收一个 CSS 选择器或一个 CSS 选择器列表，提取 CSS 选择器选中区域下的链接。
  示例 提取页面 example1.html 中<div id="bottom">元素下的链接：

```
>>> from scrapy.linkextractors import LinkExtractor
>>> le = LinkExtractor(restrict_css='div#bottom')
>>> links = le.extract_links(response1)
>>> [link.url for link in links]
['http://stackoverflow.com/tags/scrapy/info',
 'https://github.com/scrapy/scrapy']
```

- tags

  接收一个标签(字符串)或一个标签列表,提取指定标签内的链接,默认为['a', 'area']。
- attrs

  接收一个属性(字符串)或一个属性列表,提取指定属性内的链接,默认为['href']。

  示例 提取页面 example2.html 中引用 JavaScript 文件的链接:

```
>>> from scrapy.linkextractors import LinkExtractor
>>> le = LinkExtractor(tags='script', attrs='src')
>>> links = le.extract_links(response2)
>>> [link.url for link in links]
['http://example2.com/js/app1.js',
 'http://example2.com/js/app2.js']
```

- process_value

  接收一个形如 func(value)的回调函数。如果传递了该参数,LinkExtractor 将调用该回调函数对提取的每一个链接(如 a 的 href)进行处理,回调函数正常情况下应返回一个字符串(处理结果),想要抛弃所处理的链接时,返回 None。

  示例 在页面 example2.html 中,某些 a 的 href 属性是一段 JavaScript 代码,代码中包含了链接页面的实际 url 地址,此时应对链接进行处理,提取页面 example2.html 中所有实际链接:

```
>>> import re
>>> def process(value):
...     m = re.search("javascript:goToPage\('(.*?)'", value)
...     # 如果匹配,就提取其中 url 并返回,不匹配则返回原值
...     if m:
...         value = m.group(1)
...     return value
...
>>> from scrapy.linkextractors import LinkExtractor
>>> le = LinkExtractor(process_value=process)
>>> links = le.extract_links(response2)
>>> [link.url for link in links]
['http://example2.com/home.html',
 'http://example2.com/doc.html',
 'http://example2.com/example.html']
```

到此，我们介绍完了 LinkExtractor 构造器的各个参数，实际应用时可以同时使用一个或多个参数描述提取规则，这里不再举例。

6.3 本章小结

在 Scrapy 中，可以使用 Selector 或 LinkExtractor 提取页面中的链接，本章主要介绍后一种方法，先以一个案例展示了使用 LinkExtractor 提取链接的流程，然后详细讲解如何使用 LinkExtractor 的构造器参数描述提取规则。

第 7 章

使用 Exporter 导出数据

通过之前章节的学习，大家掌握了 Scrapy 中爬取数据、封装数据、处理数据的相关技术，本章我们来学习如何将爬取到的数据以某种数据格式保存到文件中，即导出数据。

在 Scrapy 中，负责导出数据的组件被称为 Exporter（导出器），Scrapy 内部实现了多个 Exporter，每个 Exporter 实现一种数据格式的导出，支持的数据格式如下（括号中为相应的 Exporter）：

（1）JSON (JsonItemExporter)
（2）JSON lines (JsonLinesItemExporter)
（3）CSV (CsvItemExporter)
（4）XML (XmlItemExporter)
（5）Pickle (PickleItemExporter)
（6）Marshal (MarshalItemExporter)

其中，前 4 种是极为常用的文本数据格式，而后两种是 Python 特有的。在大多数情况下，使用 Scrapy 内部提供的 Exporter 就足够了，需要以其他数据格式（上述 6 种以外）导出数据时，可以自行实现 Exporter。

7.1 指定如何导出数据

在导出数据时，需向 Scrapy 爬虫提供以下信息：

- 导出文件路径。
- 导出数据格式（即选用哪个 Exporter）。

可以通过以下两种方式指定爬虫如何导出数据：

（1）通过命令行参数指定。
（2）通过配置文件指定。

7.1.1 命令行参数

在运行 scrapy crawl 命令时，可以分别使用-o 和-t 参数指定导出文件路径以及导出数据格式。

在第 1 章 example 项目中，我们使用以下命令运行爬虫：

```
$ scrapy crawl books -o books.csv
...
$ head -10 books.csv        # 查看文件开头的 10 行
name,price
A Light in the Attic,£51.77
Tipping the Velvet,£53.74
Soumission,£50.10
Sharp Objects,£47.82
Sapiens: A Brief History of Humankind,£54.23
The Requiem Red,£22.65
The Dirty Little Secrets of Getting Your Dream Job,£33.34
"The Coming Woman: A Novel Based on the Life of the Infamous Feminist, Victoria Woodhull",£17.93
The Boys in the Boat: Nine Americans and Their Epic Quest for Gold at the 1936 Berlin Olympics,£22.60
```

其中，-o books.csv 指定了导出文件的路径，在这里虽然没有使用-t 参数指定导出数据格式，但 Scrapy 爬虫通过文件后缀名推断出我们想以 csv 作为导出数据格式。同样的道理，如果将参数改为-o books.json，Scrapy 爬虫就会以 json 作为导出数据格式。

需要明确地指定导出数据格式时，使用-t 参数，例如：

```
$ scrapy crawl books -t csv   -o books1.data
...
$ scrapy crawl books -t json -o books2.data
...
$ scrapy crawl books -t xml   -o books3.data
...
```

运行以上命令后，Scrapy 爬虫会以-t 参数中的数据格式字符串（如 csv、json、xml）为键，在配置字典 FEED_EXPORTERS 中搜索 Exporter，FEED_EXPORTERS 的内容由以下两个字典的内容合并而成：

- 默认配置文件中的 FEED_EXPORTERS_BASE。
- 用户配置文件中的 FEED_EXPORTERS。

前者包含内部支持的导出数据格式，后者包含用户自定义的导出数据格式。以下是 Scrapy 源码中定义的 FEED_EXPORTERS_BASE，它位于 scrapy.settings.default_settings 模块：

```
FEED_EXPORTERS_BASE = {
    'json': 'scrapy.exporters.JsonItemExporter',
    'jsonlines': 'scrapy.exporters.JsonLinesItemExporter',
    'jl': 'scrapy.exporters.JsonLinesItemExporter',
    'csv': 'scrapy.exporters.CsvItemExporter',
    'xml': 'scrapy.exporters.XmlItemExporter',
    'marshal': 'scrapy.exporters.MarshalItemExporter',
    'pickle': 'scrapy.exporters.PickleItemExporter',
}
```

如果用户添加了新的导出数据格式（即实现了新的 Exporter），可在配置文件 settings.py 中定义 FEED_EXPORTERS，例如：

```
FEED_EXPORTERS = {'excel': 'my_project.my_exporters.ExcelItemExporter'}
```

另外，指定导出文件路径时，还可以使用%(name)s 和%(time)s 两个特殊变量：

- %(name)s：会被替换为 Spider 的名字。
- %(time)s：会被替换为文件创建时间。

请看一个例子，假设一个项目中有爬取书籍信息、游戏信息、新闻信息的 3 个 Spider，分别名为'books'、'games'、'news'。对于任意 Spider 的任意一次爬取，都可以使用

'export_data/%(name)s/%(time)s.csv'作为导出路径，Scrapy 爬虫会依据 Spider 的名字和爬取的时间点创建导出文件：

```
$ scrapy crawl books -o 'export_data/%(name)s/%(time)s.csv'
...
$ scrapy crawl games -o 'export_data/%(name)s/%(time)s.csv'
...
$ scrapy crawl news -o 'export_data/%(name)s/%(time)s.csv'
...
$ scrapy crawl books -o 'export_data/%(name)s/%(time)s.csv'
...
$ tree export_data
export_data/
├── books
│   ├── 2017-03-06T02-31-57.csv
│   └── 2017-06-07T04-45-13.csv
├── games
│   └── 2017-04-05T01-43-01.csv
└── news
    └── 2017-05-06T09-44-06.csv
```

使用命令行参数指定如何导出数据很方便，但命令行参数只能指定导出文件路径以及导出数据格式，并且每次都在命令行里输入很长的参数让人很烦躁，使用配置文件可以弥补这些不足。

7.1.2 配置文件

接下来，我们学习在配置文件中指定如何导出数据。下面依次说明几个常用选项：

- FEED_URI
  导出文件路径。

```
FEED_URI = 'export_data/%(name)s.data'
```

- FEED_FORMAT
  导出数据格式。

```
FEED_FORMAT = 'csv'
```

- FEED_EXPORT_ENCODING
  导出文件编码（默认情况下 json 文件使用数字编码，其他使用 utf-8 编码）。

FEED_EXPORT_ENCODING = 'gbk'

- FEED_EXPORT_FIELDS
  导出数据包含的字段（默认情况下导出所有字段），并指定次序。

FEED_EXPORT_FIELDS = ['name', 'author', 'price']

- FEED_EXPORTERS
  用户自定义 Exporter 字典，添加新的导出数据格式时使用。

FEED_EXPORTERS = {'excel': 'my_project.my_exporters.ExcelItemExporter'}

7.2 添加导出数据格式

在某些需求下，我们想要添加新的导出数据格式，此时需要实现新的 Exporter 类。下面先参考 Scrapy 内部的 Exporter 类是如何实现的，然后自行实现一个 Exporter。

7.2.1 源码参考

Scrapy 内部的 Exporter 类在 scrapy.exporters 模块中实现，以下是其中的代码片段：

```python
class BaseItemExporter(object):
    def __init__(self, **kwargs):
        self._configure(kwargs)
    def _configure(self, options, dont_fail=False):
        self.encoding = options.pop('encoding', None)
        self.fields_to_export = options.pop('fields_to_export', None)
        self.export_empty_fields = options.pop('export_empty_fields', False)
        if not dont_fail and options:
            raise TypeError("Unexpected options: %s" % ', '.join(options.keys()))
    def export_item(self, item):
        raise NotImplementedError
    def serialize_field(self, field, name, value):
        serializer = field.get('serializer', lambda x: x)
        return serializer(value)
    def start_exporting(self):
        pass
    def finish_exporting(self):
        pass
```

```python
    def _get_serialized_fields(self, item, default_value=None, include_empty=None):
        """Return the fields to export as an iterable of tuples
        (name, serialized_value)
        """
        if include_empty is None:
            include_empty = self.export_empty_fields
        if self.fields_to_export is None:
            if include_empty and not isinstance(item, dict):
                field_iter = six.iterkeys(item.fields)
            else:
                field_iter = six.iterkeys(item)
        else:
            if include_empty:
                field_iter = self.fields_to_export
            else:
                field_iter = (x for x in self.fields_to_export if x in item)
        for field_name in field_iter:
            if field_name in item:
                field = {} if isinstance(item, dict) else item.fields[field_name]
                value = self.serialize_field(field, field_name, item[field_name])
            else:
                value = default_value
            yield field_name, value
# json
class JsonItemExporter(BaseItemExporter):
    def __init__(self, file, **kwargs):
        self._configure(kwargs, dont_fail=True)
        self.file = file
        kwargs.setdefault('ensure_ascii', not self.encoding)
        self.encoder = ScrapyJSONEncoder(**kwargs)
        self.first_item = True
    def start_exporting(self):
        self.file.write(b"[\n")
    def finish_exporting(self):
        self.file.write(b"\n]")
    def export_item(self, item):
        if self.first_item:
            self.first_item = False
```

```
            else:
                self.file.write(b',\n')
            itemdict = dict(self._get_serialized_fields(item))
            data = self.encoder.encode(itemdict)
            self.file.write(to_bytes(data, self.encoding))

# json lines
class JsonLinesItemExporter(BaseItemExporter):
    ...
# xml
class XmlItemExporter(BaseItemExporter):
    ...
# csv
class CsvItemExporter(BaseItemExporter):
    ...
...
```

其中的每一个 Exporter 都是 BaseItemExporter 的一个子类，BaseItemExporter 定义了一些抽象接口待子类实现：

- export_item(self, item)
  负责导出爬取到的每一项数据，参数 item 为一项爬取到的数据，每个子类必须实现该方法。
- start_exporting(self)
  在导出开始时被调用，可在该方法中执行某些初始化工作。
- finish_exporting(self)
  在导出完成时被调用，可在该方法中执行某些清理工作。

以 JsonItemExporter 为例，其实现非常简单：

- 为了使最终导出结果是一个 json 中的列表，在 start_exporting 和 finish_exporting 方法中分别向文件写入 b"[\n, b"]\n"。
- 在 export_item 方法中，调用 self.encoder.encode 方法将一项数据转换成 json 串（具体细节不再赘述），然后写入文件。

7.2.2 实现 Exporter

接下来，我们参照 JsonItemExporter 的源码，在第 1 章 example 项目中实现一个能将数据以 Excel 格式导出的 Exporter。

在项目中创建一个 my_exporters.py（与 settings.py 同级目录），在其中实现 ExcelItemExporter，代码如下：

```python
from scrapy.exporters import BaseItemExporter
import xlwt
class ExcelItemExporter(BaseItemExporter):
    def __init__(self, file, **kwargs):
        self._configure(kwargs)
        self.file = file
        self.wbook = xlwt.Workbook()
        self.wsheet = self.wbook.add_sheet('scrapy')
        self.row = 0
    def finish_exporting(self):
        self.wbook.save(self.file)

    def export_item(self, item):
        fields = self._get_serialized_fields(item)
        for col, v in enumerate(x for _, x in fields):
            self.wsheet.write(self.row, col, v)
        self.row += 1
```

解释上述代码如下：

- 这里使用第三方库 xlwt 将数据写入 Excel 文件。
- 在构造器方法中创建 Workbook 对象和 Worksheet 对象，并初始化用来记录写入行坐标的 self.row。
- 在 export_item 方法中调用基类的 _get_serialized_fields 方法，获得 item 所有字段的迭代器，然后调用 self.wsheet.write 方法将各字段写入 Excel 表格。
- finish_exporting 方法在所有数据都被写入 Excel 表格后被调用，在该方法中调用 self.wbook.save 方法将 Excel 表格写入 Excel 文件。

完成 ExcelItemExporter 后，在配置文件 settings.py 中添加如下代码：

```
FEED_EXPORTERS = {'excel': 'example.my_exporters.ExcelItemExporter'}
```

现在，可以使用 ExcelItemExporter 导出数据了，以 -t excel 为参数重新运行爬虫：

```
$ scrapy crawl books -t excel -o books.xls
```

图 7-1 所示为爬取完成后在 Excel 文件中观察到的结果。

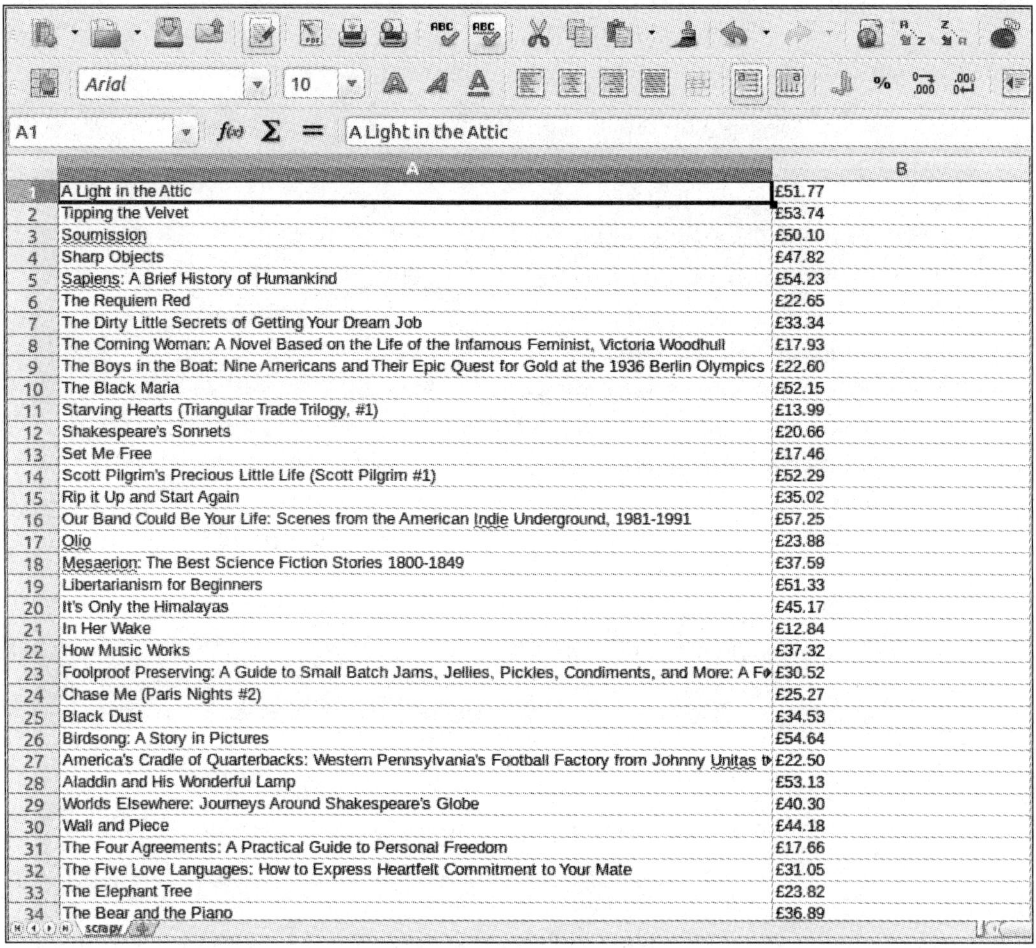

	A	B
1	A Light in the Attic	£51.77
2	Tipping the Velvet	£53.74
3	Soumission	£50.10
4	Sharp Objects	£47.82
5	Sapiens: A Brief History of Humankind	£54.23
6	The Requiem Red	£22.65
7	The Dirty Little Secrets of Getting Your Dream Job	£33.34
8	The Coming Woman: A Novel Based on the Life of the Infamous Feminist, Victoria Woodhull	£17.93
9	The Boys in the Boat: Nine Americans and Their Epic Quest for Gold at the 1936 Berlin Olympics	£22.60
10	The Black Maria	£52.15
11	Starving Hearts (Triangular Trade Trilogy, #1)	£13.99
12	Shakespeare's Sonnets	£20.66
13	Set Me Free	£17.46
14	Scott Pilgrim's Precious Little Life (Scott Pilgrim #1)	£52.29
15	Rip it Up and Start Again	£35.02
16	Our Band Could Be Your Life: Scenes from the American Indie Underground, 1981-1991	£57.25
17	Olio	£23.88
18	Mesaerion: The Best Science Fiction Stories 1800-1849	£37.59
19	Libertarianism for Beginners	£51.33
20	It's Only the Himalayas	£45.17
21	In Her Wake	£12.84
22	How Music Works	£37.32
23	Foolproof Preserving: A Guide to Small Batch Jams, Jellies, Pickles, Condiments, and More: A F	£30.52
24	Chase Me (Paris Nights #2)	£25.27
25	Black Dust	£34.53
26	Birdsong: A Story in Pictures	£54.64
27	America's Cradle of Quarterbacks: Western Pennsylvania's Football Factory from Johnny Unitas t	£22.50
28	Aladdin and His Wonderful Lamp	£53.13
29	Worlds Elsewhere: Journeys Around Shakespeare's Globe	£40.30
30	Wall and Piece	£44.18
31	The Four Agreements: A Practical Guide to Personal Freedom	£17.66
32	The Five Love Languages: How to Express Heartfelt Commitment to Your Mate	£31.05
33	The Elephant Tree	£23.82
34	The Bear and the Piano	£36.89

图 7-1

如上所示，我们成功地使用 ExcelItemExporter 将爬取到的数据存入了 Excel 文件中。

7.3　本章小结

本章学习了在 Scrapy 中如何使用 Exporter 将爬取到的数据导出到文件，首先介绍使用命令行参数以及配置文件指定如何导出数据的方法，然后参考 Scrapy 内部 Exporter 的源码实现了一个能将数据导出到 Excel 文件的 Exporter。

第 8 章

项目练习

通过之前章节的学习，大家掌握了编写 Scrapy 爬虫的基础知识，这一章我们运用之前所学进行实战项目练习。

在第 1 章的 example 项目中，我们爬取了 http://books.toscrape.com 网站中的书籍信息，但仅从每一个书籍列表页面爬取了书的名字和价格信息，如图 8-1 所示。

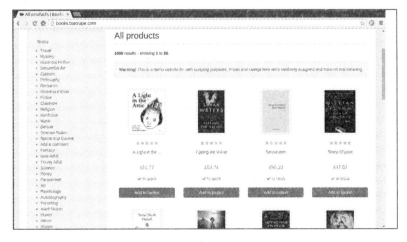

图 8-1

通常，实际应用需求并不会这么简单，可能需要获取每本书的更多信息，在具体一本书的页面中可以找到更多的信息，点击第一本书的链接，将看到如图 8-2 所示的页面。

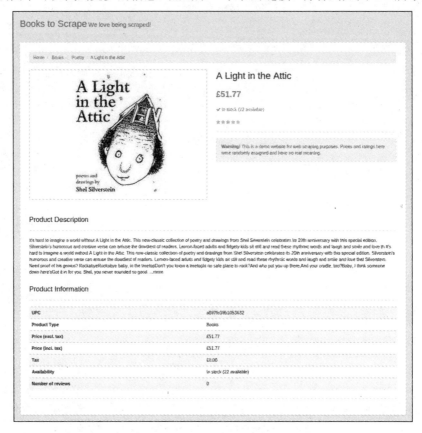

图 8-2

如上所示，在一本书的页面中可以获取以下信息：

- 书名 ✓
- 价格 ✓
- 评价等级 ✓
- 书籍简介
- 产品编码 ✓
- 产品类型
- 税价
- 库存量 ✓
- 评价数量 ✓

下面我们新建一个 Scrapy 项目，爬取每一本书更多的信息（只爬取其中打对号的信息）。

8.1 项目需求

下面爬取 http://books.toscrape.com 网站中的书籍信息。

（1）其中每一本书的信息包括：

- 书名
- 价格
- 评价等级
- 产品编码
- 库存量
- 评价数量

（2）将爬取的结果保存到 csv 文件中。

8.2 页面分析

首先，我们对一本书的页面进行分析。在进行页面分析时，除了之前使用过的 Chrome 开发者工具外，另一个常用的工具是 scrapy shell <URL>命令，它使用户可以在交互式命令行下操作一个 Scrapy 爬虫，通常我们利用该工具进行前期爬取实验，从而提高开发效率。

接下来分析第一本书的页面，以页面的 url 地址为参数运行 scrapy shell 命令：

```
$ scrapy shell http://books.toscrape.com/catalogue/a-light-in-the-attic_1000/index.html
scrapy shell http://books.toscrape.com/catalogue/a-light-in-the-attic_1000/index.html
2017-03-03 09:17:01 [scrapy] INFO: Scrapy 1.3.3 started (bot: scrapybot)
2017-03-03 09:17:01 [scrapy] INFO: Overridden settings: {'LOGSTATS_INTERVAL': 0, 'DUPEFILTER_CLASS': 'scrapy.dupefilters.BaseDupeFilter'}
2017-03-03 09:17:01 [scrapy] INFO: Enabled extensions:
['scrapy.extensions.corestats.CoreStats',
 'scrapy.extensions.telnet.TelnetConsole']
2017-03-03 09:17:01 [scrapy] INFO: Enabled downloader middlewares:
```

```
['scrapy.downloadermiddlewares.httpauth.HttpAuthMiddleware',
 'scrapy.downloadermiddlewares.downloadtimeout.DownloadTimeoutMiddleware',
 'scrapy.downloadermiddlewares.defaultheaders.DefaultHeadersMiddleware',
 'scrapy.downloadermiddlewares.useragent.UserAgentMiddleware',
 'scrapy.downloadermiddlewares.retry.RetryMiddleware',
 'scrapy.downloadermiddlewares.redirect.MetaRefreshMiddleware',
 'scrapy.downloadermiddlewares.httpcompression.HttpCompressionMiddleware',
 'scrapy.downloadermiddlewares.redirect.RedirectMiddleware',
 'scrapy.downloadermiddlewares.cookies.CookiesMiddleware',
 'scrapy.downloadermiddlewares.chunked.ChunkedTransferMiddleware',
 'scrapy.downloadermiddlewares.stats.DownloaderStats']
2017-03-03 09:17:01 [scrapy] INFO: Enabled spider middlewares:
['scrapy.spidermiddlewares.httperror.HttpErrorMiddleware',
 'scrapy.spidermiddlewares.offsite.OffsiteMiddleware',
 'scrapy.spidermiddlewares.referer.RefererMiddleware',
 'scrapy.spidermiddlewares.urllength.UrlLengthMiddleware',
 'scrapy.spidermiddlewares.depth.DepthMiddleware']
2017-03-03 09:17:01 [scrapy] INFO: Enabled item pipelines:
[]
2017-03-03 09:17:01 [scrapy] DEBUG: Telnet console listening on 127.0.0.1:6024
2017-03-03 09:17:01 [scrapy] INFO: Spider opened
2017-03-03 09:17:01 [scrapy] DEBUG: Crawled (200)   (referer: None)
2017-03-03 09:17:02 [traitlets] DEBUG: Using default logger
2017-03-03 09:17:02 [traitlets] DEBUG: Using default logger
[s] Available Scrapy objects:
[s]   scrapy        scrapy module (contains scrapy.Request, scrapy.Selector, etc)
[s]   crawler
[s]   item          {}
[s]   request
[s]   response      <200 http://books.toscrape.com/catalogue/a-light-in-the-attic_1000/index.html>
[s]   settings
[s]   spider
[s] Useful shortcuts:
[s]   shelp()              Shell help (print this help)
[s]   fetch(req_or_url)    Fetch request (or URL) and update local objects
[s]   view(response)       View response in a browser
>>>
```

运行这条命令后，scrapy shell 会使用 url 参数构造一个 Request 对象，并提交给 Scrapy 引擎，页面下载完成后，程序进入一个 python shell 当中，在此环境中已经创建好了一些变量（对象和函数），以下几个最为常用：

- request
  最近一次下载对应的 Request 对象。
- response
  最近一次下载对应的 Response 对象。
- fetch(req_or_url)
  该函数用于下载页面，可传入一个 Request 对象或 url 字符串，调用后会更新变量 request 和 response。
- view(response)
  该函数用于在浏览器中显示 response 中的页面。

接下来，在 scrapy shell 中调用 view 函数，在浏览器中显示 response 所包含的页面：

```
>>> view(response)
```

可能在很多时候，使用 view 函数打开的页面和在浏览器直接输入 url 打开的页面看起来是一样的，但需要知道的是，前者是由 Scrapy 爬虫下载的页面，而后者是由浏览器下载的页面，有时它们是不同的。在进行页面分析时，使用 view 函数更加可靠。下面使用 Chrome 审查元素工具分析页面，如图 8-3 所示。

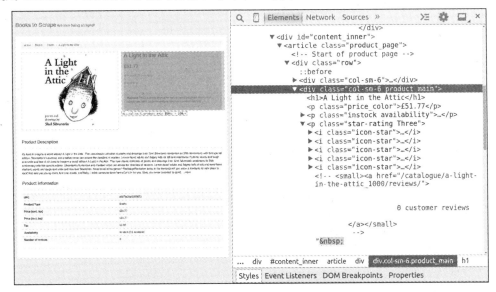

图 8-3

从图 8-3 中看出，我们可在<div class="col-sm-6 product_main">中提取书名、价格、评价等级，在 scrapy shell 中尝试提取这些信息，如图 8-4 所示。

```
>>> sel = response.css('div.product_main')
>>> sel.xpath('./h1/text()').extract_first()
'A Light in the Attic'
>>> sel.css('p.price_color::text').extract_first()
'£51.77'
>>> sel.css('p.star-rating::attr(class)').re_first('star-rating ([A-Za-z]+)')
'Three'
```

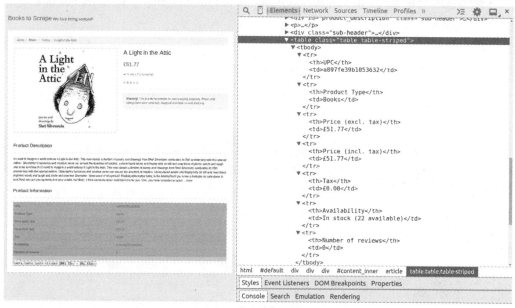

图 8-4

另外，可在页面下端位置的<table class="table table-striped">中提取产品编码、库存量、评价数量，在 scrapy shell 中尝试提取这些信息：

```
>>> sel = response.css('table.table.table-striped')
>>> sel.xpath('(.//tr)[1]/td/text()').extract_first()
'a897fe39b1053632'
>>> sel.xpath('(.//tr)[last()-1]/td/text()').re_first('\((\d+) available\)')
'22'
>>> sel.xpath('(.//tr)[last()]/td/text()').extract_first()
'0'
```

分析完书籍页面后，接着分析如何在书籍列表页面中提取每一个书籍页面的链接。在 scrapy shell 中，先调用 fetch 函数下载第一个书籍列表页面（http://books.toscrape.com/），下载完成后再调用 view 函数在浏览器中查看页面，如图 8-5 所示。

```
>>> fetch('http://books.toscrape.com/')
[scrapy] DEBUG: Crawled (200) <GET http://books.toscrape.com/> (referer: None)
>>> view(response)
```

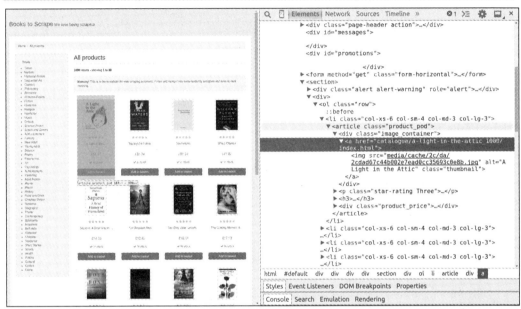

图 8-5

每个书籍页面的链接可以在每个<article class="product_pod">中找到，在 scrapy shell 中使用 LinkExtractor 提取这些链接：

```
>>> from scrapy.linkextractors import LinkExtractor
>>> le = LinkExtractor(restrict_css='article.product_pod')
>>> le.extract_links(response)
[Link(url='http://books.toscrape.com/catalogue/a-light-in-the-attic_1000/index.html', text='', fragment='', nofollow=False),
 Link(url='http://books.toscrape.com/catalogue/tipping-the-velvet_999/index.html', text='', fragment='', nofollow=False),
 Link(url='http://books.toscrape.com/catalogue/soumission_998/index.html', text='', fragment='', nofollow=False),
```

　　　　Link(url='http://books.toscrape.com/catalogue/sharp-objects_997/index.html', text='', fragment='', nofollow=False),
　　　　Link(url='http://books.toscrape.com/catalogue/sapiens-a-brief-history-of-humankind_996/index.html', text='', fragment='', nofollow=False),
　　　　Link(url='http://books.toscrape.com/catalogue/the-requiem-red_995/index.html', text='', fragment='', nofollow=False),
　　　　Link(url='http://books.toscrape.com/catalogue/the-dirty-little-secrets-of-getting-your-dream-job_994/index.html', text='', fragment='', nofollow=False),
　　　　Link(url='http://books.toscrape.com/catalogue/the-coming-woman-a-novel-based-on-the-life-of-the-infamous-feminist-victoria-woodhull_993/index.html', text='', fragment='', nofollow=False),
　　　　Link(url='http://books.toscrape.com/catalogue/the-boys-in-the-boat-nine-americans-and-their-epic-quest-for-gold-at-the-1936-berlin-olympics_992/index.html', text='', fragment='', nofollow=False),
　　　　Link(url='http://books.toscrape.com/catalogue/the-black-maria_991/index.html', text='', fragment='', nofollow=False),
　　　　Link(url='http://books.toscrape.com/catalogue/starving-hearts-triangular-trade-trilogy-1_990/index.html', text='', fragment='', nofollow=False),
　　　　Link(url='http://books.toscrape.com/catalogue/shakespeares-sonnets_989/index.html', text='', fragment='', nofollow=False),
　　　　Link(url='http://books.toscrape.com/catalogue/set-me-free_988/index.html', text='', fragment='', nofollow=False),
　　　　Link(url='http://books.toscrape.com/catalogue/scott-pilgrims-precious-little-life-scott-pilgrim-1_987/index.html', text='', fragment='', nofollow=False),
　　　　Link(url='http://books.toscrape.com/catalogue/rip-it-up-and-start-again_986/index.html', text='', fragment='', nofollow=False),
　　　　Link(url='http://books.toscrape.com/catalogue/our-band-could-be-your-life-scenes-from-the-american-indie-underground-1981-1991_985/index.html', text='', fragment='', nofollow=False),
　　　　Link(url='http://books.toscrape.com/catalogue/olio_984/index.html', text='', fragment='', nofollow=False),
　　　　Link(url='http://books.toscrape.com/catalogue/mesaerion-the-best-science-fiction-stories-1800-1849_983/index.html', text='', fragment='', nofollow=False),
　　　　Link(url='http://books.toscrape.com/catalogue/libertarianism-for-beginners_982/index.html', text='', fragment='', nofollow=False),
　　　　Link(url='http://books.toscrape.com/catalogue/its-only-the-himalayas_981/index.html', text='', fragment='', nofollow=False)]

到此，页面分析的工作已经完成了。

8.3 编码实现

首先创建一个 Scrapy 项目，取名为 toscrape_book。

```
$ scrapy startproject toscrape_book
```

通常，我们不需要手工创建 Spider 文件以及 Spider 类，可以使用 scrapy genspider <SPIDER_NAME> <DOMAIN>命令生成（根据模板）它们，该命令的两个参数分别是 Spider 的名字和所要爬取的域（网站）：

```
$ cd toscrape_book
$ scrapy genspider books books.toscrape.com
```

运行后，scrapy genspider 命令创建了文件 toscrape_book/spiders/books.py，并在其中创建了一个 BooksSpider 类，代码如下：

```
# -*- coding: utf-8 -*-
import scrapy

class BooksSpider(scrapy.Spider):
    name = "books"
    allowed_domains = ["books.toscrape.com"]
    start_urls = ['http://books.toscrape.com/']

    def parse(self, response):
        pass
```

实现 Spider 之前，先定义封装书籍信息的 Item 类，在 toscrape_book/items.py 中添加如下代码：

```
class BookItem(scrapy.Item):
    name = scrapy.Field()              # 书名
    price = scrapy.Field()             # 价格
    review_rating = scrapy.Field()     # 评价等级，1~5 星
    review_num = scrapy.Field()        # 评价数量
    upc = scrapy.Field()               # 产品编码
    stock = scrapy.Field()             # 库存量
```

接下来，按以下 5 步完成 BooksSpider。

步骤 01 继承 Spider 创建 BooksSpider 类（已完成）。

步骤02 为 Spider 取名（已完成）。
步骤03 指定起始爬取点（已完成）。
步骤04 实现书籍列表页面的解析函数。
步骤05 实现书籍页面的解析函数。

其中前 3 步已经由 scrapy genspider 命令帮我们完成，不需做任何修改。

第 4 步和第 5 步的工作是实现两个页面解析函数，因为起始爬取点是一个书籍列表页面，我们就将 parse 方法作为书籍列表页面的解析函数，另外，还需要添加一个 parse_book 方法作为书籍页面的解析函数，代码如下：

```python
class BooksSpider(scrapy.Spider):
    name = "books"
    allowed_domains = ["books.toscrape.com"]
    start_urls = ['http://books.toscrape.com/']

    # 书籍列表页面的解析函数
    def parse(self, response):
        pass

    # 书籍页面的解析函数
    def parse_book(self, reponse):
        pass
```

先来完成第 4 步，实现书籍列表页面的解析函数（parse 方法），需要完成以下两个任务：

（1）提取页面中每一个书籍页面的链接，用它们构造 Request 对象并提交。
（2）提取页面中下一个书籍列表页面的链接，用其构造 Request 对象并提交。

提取链接的具体细节在页面分析时已经讨论过，实现代码如下：

```python
class BooksSpider(scrapy.Spider):
    name = "books"
    allowed_domains = ["books.toscrape.com"]
    start_urls = ['http://books.toscrape.com/']

    # 书籍列表页面的解析函数
    def parse(self, response):
```

```
        # 提取书籍列表页面中每本书的链接
        le = LinkExtractor(restrict_css='article.product_pod h3')
        for link in le.extract_links(response):
            yield scrapy.Request(link.url, callback=self.parse_book)

        # 提取"下一页"的链接
        le = LinkExtractor(restrict_css='ul.pager li.next')
        links = le.extract_links(response)
        if links:
            next_url = links[0].url
            yield scrapy.Request(next_url, callback=self.parse)

    # 书籍页面的解析函数
    def parse_book(self, response):
        pass
```

最后完成第 5 步，实现书籍页面的解析函数（parse_book 方法），只需提取书籍信息存入 BookItem 对象即可。同样，提取书籍信息的细节也在页面分析时讨论过，最终完成代码如下：

```
import scrapy
from scrapy.linkextractors import LinkExtractor
from ..items import BookItem

class BooksSpider(scrapy.Spider):
    name = "books"
    allowed_domains = ["books.toscrape.com"]
    start_urls = ['http://books.toscrape.com/']

    def parse(self, response):
        le = LinkExtractor(restrict_css='article.product_pod h3')
        for link in le.extract_links(response):
            yield scrapy.Request(link.url, callback=self.parse_book)

        le = LinkExtractor(restrict_css='ul.pager li.next')
        links = le.extract_links(response)
        if links:
            next_url = links[0].url
            yield scrapy.Request(next_url, callback=self.parse)
```

```python
def parse_book(self, response):
    book = BookItem()
    sel = response.css('div.product_main')
    book['name'] = sel.xpath('./h1/text()').extract_first()
    book['price'] = sel.css('p.price_color::text').extract_first()
    book['review_rating'] = sel.css('p.star-rating::attr(class)')\
                            .re_first('star-rating ([A-Za-z]+)')

    sel = response.css('table.table.table-striped')
    book['upc'] = sel.xpath('(.//tr)[1]/td/text()').extract_first()
    book['stock'] = sel.xpath('(.//tr)[last()-1]/td/text()')\
                    .re_first('\((\d+) available\)')
    book['review_num'] = sel.xpath('(.//tr)[last()]/td/text()').extract_first()

    yield book
```

完成代码后，运行爬虫并观察结果：

```
$ scrapy crawl books -o books.csv --nolog
$ cat -n books.csv
    1  name,stock,price,review_num,review_rating,upc
    2  Scott Pilgrim's Precious Little Life,19,£52.29,0,Five,3b1c02bac2a429e6
    3  It's Only the Himalayas,19,£45.17,0,Two,a22124811bfa8350
    4  Olio,19,£23.88,0,One,feb7cc7701ecf901
    5  Rip it Up and Start Again,19,£35.02,0,Five,a34ba96d4081e6a4

    ... 省略中间输出 ...

  999  Bright Lines,1,£39.07,0,Five,230ac636ea0ea415
 1000  Jurassic Park (Jurassic Park #1),3,£44.97,0,One,a0dd11f6abc421ec
 1001  Into the Wild,3,£56.70,0,Five,a7c3f1010d64799a
```

从以上结果中看出，我们成功地爬取了网站中 1000 本书的详细信息，但也有让人不满意的地方，比如 csv 文件中各列的次序是随机的，看起来比较混乱，可在配置文件 settings.py 中使用 FEED_EXPORT_FIELDS 指定各列的次序：

```
FEED_EXPORT_FIELDS = ['upc', 'name', 'price', 'stock', 'review_rating', 'review_num']
```

另外，结果中评价等级字段的值是 One、Two、Three……这样的单词，而不是阿拉伯数字，阅读起来不是很直观。下面实现一个 Item Pipeline，将评价等级字段由单词映射到数字（或许这样简单的需求使用 Item Pipeline 有点大材小用，主要目的是带领大家复习之前所学的知识）。在 pipelines.py 中实现 BookPipeline，代码如下：

```python
class BookPipeline(object):
    review_rating_map = {
        'One':   1,
        'Two':   2,
        'Three': 3,
        'Four':  4,
        'Five':  5,
    }

    def process_item(self, item, spider):
        rating = item.get('review_rating')
        if rating:
            item['review_rating'] = self.review_rating_map[rating]

        return item
```

在配置文件 settings.py 中启用 BookPipeline：

```python
ITEM_PIPELINES = {
    'toscrape_book.pipelines.BookPipeline': 300,
}
```

重新运行爬虫，并观察结果：

```
$ scrapy crawl books -o books.csv
...
$ cat -n books.csv
     1  upc,name,price,stock,review_rating,review_num
     2  a897fe39b1053632,A Light in the Attic,£51.77,22,3,0
     3  3b1c02bac2a429e6,Scott Pilgrim's Precious Little Life,£52.29,19,5,0
     4  a22124811bfa8350,It's Only the Himalayas,£45.17,19,2,0
     5  feb7cc7701ecf901,Olio,£23.88,19,1,0

     ... 省略中间输出 ...
```

```
999    91eb9605998a7c03,"The Sandman, Vol. 3: Dream Country",£55.55,3,5,0
1000   f06039c29b5891fa,The Silkworm (Cormoran Strike #2),£23.05,3,5,0
1001   476c7972e9b41891,The Last Painting of Sara de Vos,£55.55,3,2,0
```

此时,各字段已按指定次序排列,并且评价等级字段的值是我们所期望的阿拉伯数字。

到此为止,整个项目完成了。

8.4 本章小结

本章是基础篇的最后一章,通过一个 Scrapy 爬虫项目复习了之前章节所学的知识,现在,大家已经能够编写一个一般任务的 Scrapy 爬虫了,可以通过更多的实战项目进行练习,在后面的章节中,我们将会学习一些高级话题。

第 9 章

下载文件和图片

在之前的章节中,我们学习了从网页中爬取信息的方法,这只是爬虫最典型的一种应用,除此之外,下载文件也是实际应用中很常见的一种需求,例如使用爬虫爬取网站中的图片、视频、WORD 文档、PDF 文件、压缩包等。本章来学习在 Scrapy 中如何下载文件和图片。

9.1 FilesPipeline 和 ImagesPipeline

Scrapy 框架内部提供了两个 Item Pipeline,专门用于下载文件和图片:

- FilesPipeline
- ImagesPipeline

我们可以将这两个 Item Pipeline 看作特殊的下载器,用户使用时只需要通过 item 的一个特殊字段将要下载文件或图片的 url 传递给它们,它们会自动将文件或图片下载到本地,并将下载结果信息存入 item 的另一个特殊字段,以便用户在导出文件中查阅。下面详细介绍如何使用它们。

9.1.1 FilesPipeline 使用说明

通过一个简单的例子讲解 FilesPipeline 的使用，在如下页面中可以下载多本 PDF 格式的小说：

```
<html>
  <body>
    ...
    <a href='/book/sg.pdf'>下载《三国演义》</a>
    <a href='/book/shz.pdf'>下载《水浒传》</a>
    <a href='/book/hlm.pdf'>下载《红楼梦》</a>
    <a href='/book/xyj.pdf'>下载《西游记》</a>
    ...
  </body>
</html>
```

使用 FilesPipeline 下载页面中所有 PDF 文件，可按以下步骤进行：

步骤 01 在配置文件 settings.py 中启用 FilesPipeline，通常将其置于其他 Item Pipeline 之前：

```
ITEM_PIPELINES = {'scrapy.pipelines.files.FilesPipeline': 1}
```

步骤 02 在配置文件 settings.py 中，使用 FILES_STORE 指定文件下载目录，如：

```
FILES_STORE = '/home/liushuo/Download/scrapy'
```

步骤 03 在 Spider 解析一个包含文件下载链接的页面时，将所有需要下载文件的 url 地址收集到一个列表，赋给 item 的 file_urls 字段（item['file_urls']）。FilesPipeline 在处理每一项 item 时，会读取 item['file_urls']，对其中每一个 url 进行下载，Spider 示例代码如下：

```
class DownloadBookSpider(scrapy.Spider):
    ...

    def parse(response):
        item = {}
        # 下载列表
        item['file_urls'] = []
```

```
            for url in response.xpath('//a/@href').extract():
                download_url = response.urljoin(url)
                # 将 url 填入下载列表
                item['file_urls'].append(download_url)

            yield item
```

当 FilesPipeline 下载完 item['file_urls'] 中的所有文件后，会将各文件的下载结果信息收集到另一个列表，赋给 item 的 files 字段（item['files']）。下载结果信息包括以下内容：

- Path 文件下载到本地的路径（相对于 FILES_STORE 的相对路径）。
- Checksum 文件的校验和。
- url 文件的 url 地址。

9.1.2 ImagesPipeline 使用说明

图片也是文件，所以下载图片本质上也是下载文件，ImagesPipeline 是 FilesPipeline 的子类，使用上和 FilesPipeline 大同小异，只是在所使用的 item 字段和配置选项上略有差别，如表 9-1 所示。

表 9-1 ImagesPipeline和FilesPipeline

	FilesPipeline	ImagesPipeline
导入路径	scrapy.pipelines.files.FilesPipeline	scrapy.pipelines.images.ImagesPipeline
Item 字段	file_urls, files	image_urls, images
下载目录	FILES_STORE	IMAGES_STORE

ImagesPipleline 在 FilesPipleline 的基础上针对图片增加了一些特有的功能：

- 为图片生成缩略图
  开启该功能，只需在配置文件 settings.py 中设置 IMAGES_THUMBS，它是一个字典，每一项的值是缩略图的尺寸，代码如下：

```
IMAGES_THUMBS = {
    'small': (50, 50),
    'big': (270, 270),
}
```

开启该功能后，下载一张图片时，本地会出现3张图片（1张原图片，2张缩略图），路径如下：

[IMAGES_STORE]/full/63bbfea82b8880ed33cdb762aa11fab722a90a24.jpg
[IMAGES_STORE]/thumbs/small/63bbfea82b8880ed33cdb762aa11fab722a90a24.jpg
[IMAGES_STORE]/thumbs/big/63bbfea82b8880ed33cdb762aa11fab722a90a24.jpg

- 过滤掉尺寸过小的图片

  开启该功能，需在配置文件 settings.py 中设置 IMAGES_MIN_WIDTH 和 IMAGES_MIN_HEIGHT，它们分别指定图片最小的宽和高，代码如下：

IMAGES_MIN_WIDTH = 110
IMAGES_MIN_HEIGHT = 110

开启该功能后，如果下载了一张 105×200 的图片，该图片就会被抛弃掉，因为它的宽度不符合标准。

9.2 项目实战：爬取 matplotlib 例子源码文件

下面我们来完成一个使用 FilesPipeline 下载文件的实战项目。matplotlib 是一个非常著名的 Python 绘图库，广泛应用于科学计算和数据分析等领域。在 matplotlib 网站上提供了许多应用例子代码，在浏览器中访问 http://matplotlib.org/examples/index.html，可看到图 9-1 所示的例子列表页面。

其中有几百个例子，被分成多个类别，单击第一个例子，进入其页面，如图 9-2 所示。

用户可以在每个例子页面中阅读源码，也可以点击页面中的 source code 按钮下载源码文件。如果我们想把所有例子的源码文件都下载到本地，可以编写一个爬虫程序完成这个任务。

9.2.1 项目需求

下载 http://matplotlib.org 网站中所有例子的源码文件到本地。

第 9 章 下载文件和图片

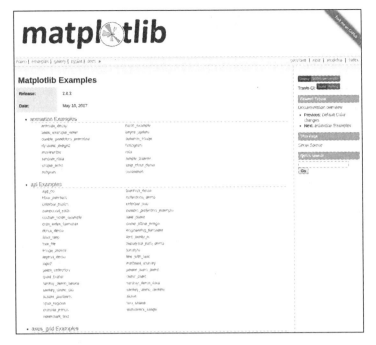

图 9-1

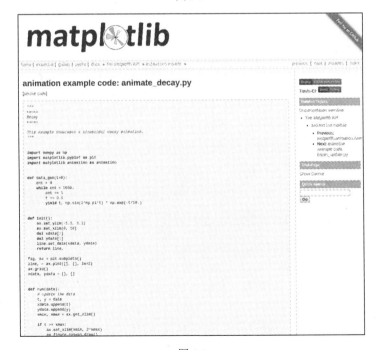

图 9-2

9.2.2 页面分析

先来看如何在例子列表页面 http://matplotlib.org/examples/index.html 中获取所有例子页面的链接。使用 scrapy shell 命令下载页面，然后调用 view 函数在浏览器中查看页面，如图 9-3 所示。

```
$ scrapy shell http://matplotlib.org/examples/index.html
...
>>> view(response)
```

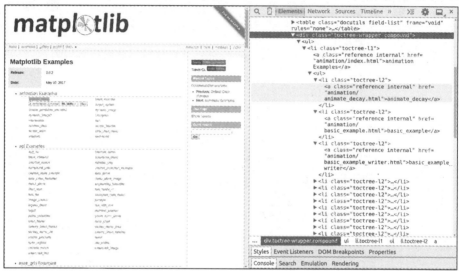

图 9-3

观察发现，所有例子页面的链接都在<div class="toctree-wrapper compound">下的每一个<li class="toctree-l2">中，例如：

<a class="reference internal" href="animation/animate_decay.html">animate_decay</a>

使用 LinkExtractor 提取所有例子页面的链接，代码如下：

```
>>> from scrapy.linkextractors import LinkExtractor
>>> le = LinkExtractor(restrict_css='div.toctree-wrapper.compound li.toctree-l2')
>>> links = le.extract_links(response)
>>> [link.url for link in links]
['http://matplotlib.org/examples/animation/animate_decay.html',
 'http://matplotlib.org/examples/animation/basic_example.html',
 'http://matplotlib.org/examples/animation/basic_example_writer.html',
```

```
'http://matplotlib.org/examples/animation/bayes_update.html',
'http://matplotlib.org/examples/animation/double_pendulum_animated.html',
'http://matplotlib.org/examples/animation/dynamic_image.html',
'http://matplotlib.org/examples/animation/dynamic_image2.html',
'http://matplotlib.org/examples/animation/histogram.html',
'http://matplotlib.org/examples/animation/moviewriter.html',
'http://matplotlib.org/examples/animation/rain.html',
'http://matplotlib.org/examples/animation/random_data.html',
'http://matplotlib.org/examples/animation/simple_3danim.html',
'http://matplotlib.org/examples/animation/simple_anim.html',
'http://matplotlib.org/examples/animation/strip_chart_demo.html',
'http://matplotlib.org/examples/animation/subplots.html',
'http://matplotlib.org/examples/animation/unchained.html',
'http://matplotlib.org/examples/api/agg_oo.html',
'http://matplotlib.org/examples/api/barchart_demo.html',
'http://matplotlib.org/examples/api/bbox_intersect.html',
...
'http://matplotlib.org/examples/user_interfaces/svg_tooltip.html',
'http://matplotlib.org/examples/user_interfaces/toolmanager.html',
'http://matplotlib.org/examples/user_interfaces/wxcursor_demo.html',
'http://matplotlib.org/examples/widgets/buttons.html',
'http://matplotlib.org/examples/widgets/check_buttons.html',
'http://matplotlib.org/examples/widgets/cursor.html',
'http://matplotlib.org/examples/widgets/lasso_selector_demo.html',
'http://matplotlib.org/examples/widgets/menu.html',
'http://matplotlib.org/examples/widgets/multicursor.html',
'http://matplotlib.org/examples/widgets/radio_buttons.html',
'http://matplotlib.org/examples/widgets/rectangle_selector.html',
'http://matplotlib.org/examples/widgets/slider_demo.html',
'http://matplotlib.org/examples/widgets/span_selector.html']
>>> len(links)
507
```

例子列表页面分析完毕,总共找到了 507 个例子。

接下来分析例子页面。调用 fetch 函数下载第一个例子页面,并调用 view 函数在浏览器中查看页面,如图 9-4 所示。

```
>>> fetch('http://matplotlib.org/examples/animation/animate_decay.html')
...
```

```
>>> view(response)
```

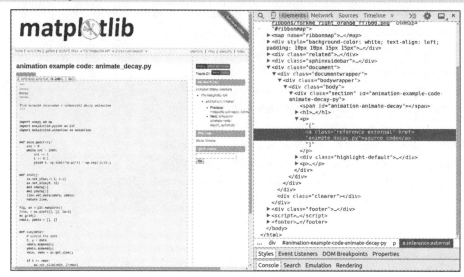

图 9-4

在一个例子页面中，例子源码文件的下载地址可在<a class="reference external">中找到：

```
>>> href = response.css('a.reference.external::attr(href)').extract_first()
>>> href
'animate_decay.py'
>>> response.urljoin(href)
'http://matplotlib.org/examples/animation/animate_decay.py'
```

到此，页面分析的工作完成了。

9.2.3 编码实现

接下来，我们按以下 4 步完成该项目：

（1）创建 Scrapy 项目，并使用 scrapy genspider 命令创建 Spider。
（2）在配置文件中启用 FilesPipeline，并指定文件下载目录。
（3）实现 ExampleItem（可选）。
（4）实现 ExamplesSpider。

步骤 01 首先创建 Scrapy 项目，取名为 matplotlib_examples，再使用 scrapy genspider 命令创建 Spider：

```
$ scrapy startproject matplotlib_examples
$ cd matplotlib_examples
$ scrapy genspider examples matplotlib.org
```

步骤 02 在配置文件 settings.py 中启用 FilesPipeline，并指定文件下载目录，代码如下：

```
ITEM_PIPELINES = {
    'scrapy.pipelines.files.FilesPipeline': 1,
}
FILES_STORE = 'examples_src'
```

步骤 03 实现 ExampleItem，需定义 file_urls 和 files 两个字段，在 items.py 中完成如下代码：

```
class ExampleItem(scrapy.Item):
    file_urls = scrapy.Field()
    files = scrapy.Field()
```

步骤 04 实现 ExamplesSpider。首先设置起始爬取点：

```
import scrapy

class ExamplesSpider(scrapy.Spider):
    name = "examples"
    allowed_domains = ["matplotlib.org"]
    start_urls = ['http://matplotlib.org/examples/index.html']

    def parse(self, response):
        pass
```

parse 方法是例子列表页面的解析函数，在该方法中提取每个例子页面的链接，用其构造 Request 对象并提交，提取链接的细节已在页面分析时讨论过，实现 parse 方法的代码如下：

```
import scrapy
from scrapy.linkextractors import LinkExtractor

class ExamplesSpider(scrapy.Spider):
    name = "examples"
    allowed_domains = ["matplotlib.org"]
    start_urls = ['http://matplotlib.org/examples/index.html']
```

```python
def parse(self, response):
    le = LinkExtractor(restrict_css='div.toctree-wrapper.compound',
                       deny='/index.html$')

    print(len(le.extract_links(response)))
    for link in le.extract_links(response):
        yield scrapy.Request(link.url, callback=self.parse_example)

def parse_example(self, response):
    pass
```

上面代码中,我们将例子页面的解析函数设置为 parse_example 方法,下面来实现这个方法。例子页面中包含了例子源码文件的下载链接,在 parse_example 方法中获取源码文件的 url,将其放入一个列表,赋给 ExampleItem 的 file_urls 字段。实现 parse_example 方法的代码如下:

```python
import scrapy
from scrapy.linkextractors import LinkExtractor
from ..items import ExampleItem
class ExamplesSpider(scrapy.Spider):
    name = "examples"
    allowed_domains = ["matplotlib.org"]
    start_urls = ['http://matplotlib.org/examples/index.html']
    def parse(self, response):
        le = LinkExtractor(restrict_css='div.toctree-wrapper.compound',
                           deny='/index.html$')

        print(len(le.extract_links(response)))
        for link in le.extract_links(response):
            yield scrapy.Request(link.url, callback=self.parse_example)
    def parse_example(self, response):
        href = response.css('a.reference.external::attr(href)').extract_first()
        url = response.urljoin(href)
        example = ExampleItem()
        example['file_urls'] = [url]
        return example
```

编码完成后,运行爬虫,并观察结果:

```
$ scrapy crawl examples -o examples.json
...
```

```
$ ls
examples.json  examples_src  matplotlib_examples  scrapy.cfg
```

运行结束后，在文件 examples.json 中可以查看到文件下载结果信息：

```
$ cat examples.json
[
{"file_urls": ["http://matplotlib.org/mpl_examples/axes_grid/demo_floating_axes.py"], "files": [{"url": "http://matplotlib.org/mpl_examples/axes_grid/demo_floating_axes.py", "checksum": "502d1cd62086fb1d4de033cef2e495c0", "path": "full/d9b551310a6668ccf43871e896f2fe6e0228567d.py"}]},
{"file_urls": ["http://matplotlib.org/mpl_examples/axes_grid/demo_curvelinear_grid.py"], "files": [{"url": "http://matplotlib.org/mpl_examples/axes_grid/demo_curvelinear_grid.py", "checksum": "5cb91103f11079b40400afc0c1f4a508", "path": "full/366386c23c5b715c49801efc7f8d55d2c74252e2.py"}]},
{"file_urls": ["http://matplotlib.org/mpl_examples/axes_grid/make_room_for_ylabel_using_axesgrid.py"], "files": [{"url": "http://matplotlib.org/mpl_examples/axes_grid/make_room_for_ylabel_using_axesgrid.py", "checksum": "dcf561f97ab0905521c1957cacd2da00", "path": "full/919cbbe6d725237e3b6051f544f6109e7189b4fe.py"}]},
...省略部分内容...
{"file_urls": ["http://matplotlib.org/mpl_examples/api/custom_projection_example.py"], "files": [{"url": "http://matplotlib.org/mpl_examples/api/custom_projection_example.py", "checksum": "bde485f9d5ceb4b4cc969ef692df5eee", "path": "full/d56af342d7130ddd9dbf55c00664eae9a432bf70.py"}]},
{"file_urls": ["http://matplotlib.org/examples/animation/dynamic_image2.py"], "files": [{"url": "http://matplotlib.org/examples/animation/dynamic_image2.py", "checksum": "98b6a6021ba841ef4a2cd36c243c516d", "path": "full/fe635002562e8685583c1b35a8e11e8cde0a6321.py"}]},
{"file_urls": ["http://matplotlib.org/examples/animation/basic_example.py"], "files": [{"url": "http://matplotlib.org/examples/animation/basic_example.py", "checksum": "1d4afc0910f6abc519e6ecd32c66896a", "path": "full/083c113c1dac96bbc74adfc5b08cad68ec9c16db.py"}]}]
```

再来查看文件下载目录 exmaples_src：

```
$ tree examples_src
examples_src
└── full
    ├── 006947659b6bd35d3f5a27df2bc36f81f1170e2e.py
```

```
      ├── 00f4d142b951f0727f1d549b1d17574b5f841776.py
      ├── 0100763ca91d69e64085cf770bf2a6f9d0c0136d.py
      ├── 019bbcfcf06fb55ab46a1e52c71f56804bd06047.py
      ├── 020bc2c11d2b0768a6df86a3b9dacf7adecf442d.py
      ├── 028c0be66575ba9ab770330f7752e8e670308813.py

      ... 省略部分输出 ...

      ├── fe7e5882c83d7c7c020bf9f3ae5570172a5fd3a6.py
      ├── ffcb2f1e36fed60034a5d961674a0218857057b6.py
      ├── ffdad211ab38c4f0ecba1857d11c7507b8e43ca8.py
      └── fff6a4cf0d0261d63c79ba3606b195c8764ffc11.py

1 directory, 507 files
```

如上所示，507个源码文件被下载到了examples_src/full目录下，并且每个文件的名字都是一串长度相等的奇怪数字，这些数字是下载文件url的sha1散列值。例如，某文件url为：

```
http://matplotlib.org/mpl_examples/axes_grid/demo_floating_axes.py
```

该url的sha1散列值为：

```
d9b551310a6668ccf43871e896f2fe6e0228567d
```

那么该文件的存储路径为：

```
# [FILES_STORE]/full/[SHA1_HASH_VALUE].py
examples_src/full/d9b551310a6668ccf43871e896f2fe6e0228567d.py
```

这种命名方式可以防止重名的文件相互覆盖，但这样的文件名太不直观了，无法从文件名了解文件内容，我们期望把这些例子文件按照类别下载到不同目录下，为完成这个任务，可以写一个单独的脚本，依据examples.json文件中的信息将文件重命名，也可以修改FilesPipeline为文件命名的规则，这里采用后一种方式。

阅读FilesPipeline的源码发现，原来是其中的file_path方法决定了文件的命名，相关代码如下：

```
class FilesPipeline(MediaPipeline):
    ...
    def file_path(self, request, response=None, info=None):
        ...
```

```python
        # check if called from file_key with url as first argument
        if not isinstance(request, Request):
            _warn()
            url = request
        else:
            url = request.url

        # detect if file_key() method has been overridden
        if not hasattr(self.file_key, '_base'):
            _warn()
            return self.file_key(url)
        ## end of deprecation warning block

        media_guid = hashlib.sha1(to_bytes(url)).hexdigest()
        media_ext = os.path.splitext(url)[1]
        return 'full/%s%s' % (media_guid, media_ext)
    ...
```

现在，我们实现一个 FilesPipeline 的子类，覆写 file_path 方法来实现所期望的文件命名规则，这些源码文件 url 的最后两部分是类别和文件名，例如：

http://matplotlib.org/mpl_examples/(axes_grid/demo_floating_axes.py)

可用以上括号中的部分作为文件路径，在 pipelines.py 实现 MyFilesPipeline，代码如下：

```python
from scrapy.pipelines.files import FilesPipeline
from urllib.parse import urlparse
from os.path import basename, dirname, join

class MyFilesPipeline(FilesPipeline):
    def file_path(self, request, response=None, info=None):
        path = urlparse(request.url).path
        return join(basename(dirname(path)), basename(path))
```

修改配置文件，使用 MyFilesPipeline 替代 FilesPipeline：

```python
ITEM_PIPELINES = {
    #'scrapy.pipelines.files.FilesPipeline': 1,
    'matplotlib_examples.pipelines.MyFilesPipeline': 1,
}
```

删除之前下载的所有文件，重新运行爬虫后，再来查看 examples_src 目录：

```
$ rm -r examples_src/full
$ rm examples.json
$ scrapy crawl examples -o examples.json
...
$ tree examples_src
examples_src/
├── animation
│   ├── animate_decay.py
│   ├── basic_example.py
│   ├── basic_example_writer.py
│   ├── bayes_update.py
│   ├── double_pendulum_animated.py
│   ├── dynamic_image2.py
│   ├── dynamic_image.py
│   ├── histogram.py
│   ├── moviewriter.py
│   ├── rain.py
│   ├── random_data.py
│   ├── simple_3danim.py
│   ├── simple_anim.py
│   ├── strip_chart_demo.py
│   ├── subplots.py
│   └── unchained.py
├── api
│   ├── agg_oo.py
│   ├── barchart_demo.py
│   ├── bbox_intersect.py
│   ├── collections_demo.py
│   ├── colorbar_basics.py
│   ├── colorbar_only.py
│   ├── compound_path.py
│   ├── custom_projection_examp
...省略中间输出...
│   ├── rec_edit_gtk_simple.py
│   ├── svg_histogram.py
│   ├── svg_tooltip.py
│   ├── toolmanager.py
```

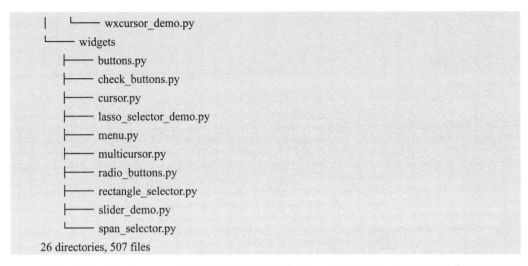

```
    │   └── wxcursor_demo.py
    └── widgets
        ├── buttons.py
        ├── check_buttons.py
        ├── cursor.py
        ├── lasso_selector_demo.py
        ├── menu.py
        ├── multicursor.py
        ├── radio_buttons.py
        ├── rectangle_selector.py
        ├── slider_demo.py
        └── span_selector.py
26 directories, 507 files
```

从上述结果看出，507 个文件按类别被下载到 26 个目录下，这正是我们所期望的。到此，文件下载的项目完成了。

9.3　项目实战：下载 360 图片

我们再来完成一个使用 ImagesPipeline 下载图片的实战项目。360 图片是一个知名的图片搜索网站，在浏览器中打开 http://image.so.com，页面如图 9-5 所示。

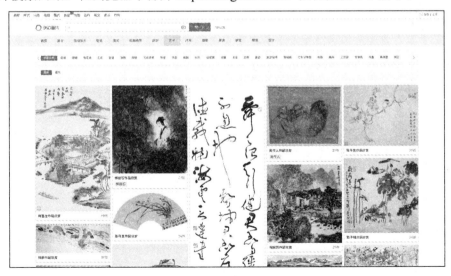

图 9-5

其中，艺术分类下有大量字画图片，我们可以编写爬虫爬取这些图片。

9.3.1 项目需求

下载360图片网站中艺术分类下的所有图片到本地。

9.3.2 页面分析

在图9-5所示的页面中向下滚动鼠标滚轮，便会有更多图片加载出来，图片加载是由JavaScript脚本完成的，在图9-6中可以看到jQuery发送的请求，其响应结果是一个json串。

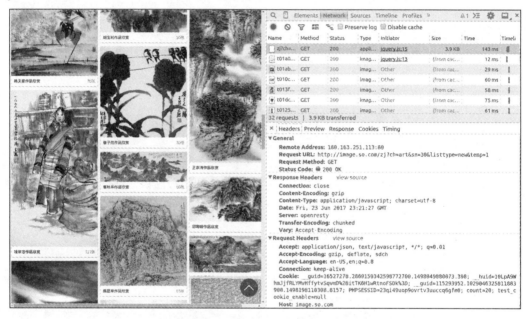

图9-6

复制图中jQuery发送请求的url，使用scrapy shell进行访问，查看响应结果的内容（json）：

```
$ scrapy shell 'http://image.so.com/zj?ch=art&sn=30&listtype=new&temp=1'
...
>>> import json
>>> res = json.loads(response.body.decode('utf8'))
>>> res
{'count': 30,
```

```
'end': False,
'lastid': 60,
'list': [{'cover_height': 942,
  'cover_imgurl': 'http://www.sinaimg.cn/dy/slidenews/26_img/2011_27/17290_50560_803601.jpg',
  'cover_width': 950,
  'dsptime': '',
  'group_title': '李正天作品欣赏',
  'grpseq': 1,
  'id': 'e4e6dbc8c5deaf2799d396569904227f',
  'imageid': '5332cbd95b1098f0e9325a16ce022a74',
  'index': 31,
  'label': '',
  'qhimg_height': 236,
  'qhimg_thumb_url': 'http://p0.so.qhimgs1.com/sdr/238__/t01ab50e7f19a03afa0.jpg',
  'qhimg_url': 'http://p0.so.qhimgs1.com/t01ab50e7f19a03afa0.jpg',
  'qhimg_width': 238,
  'tag': '新浪艺术名家人物库',
  'total_count': 70},
 {'cover_height': 1798,
  'cover_imgurl': 'http://www.sinaimg.cn/dy/slidenews/26_img/2011_15/18496_33310_603704.jpg',
  'cover_width': 950,
  'dsptime': '',
  'group_title': '崔自默作品欣赏',
  'grpseq': 1,
  'id': 'f08148a113c6c2e6104a77798d285d88',
  'imageid': 'c6662a238bb6faf9b22a335db6707fff',
  'index': 32,
  'label': '',
  'qhimg_height': 450,
  'qhimg_thumb_url': 'http://p0.so.qhmsg.com/sdr/238__/t01b187fc2ce65e29b5.jpg',
  'qhimg_url': 'http://p0.so.qhmsg.com/t01b187fc2ce65e29b5.jpg',
  'qhimg_width': 238,
  'tag': '新浪艺术名家人物库',
  'total_count': 53},
 {'cover_height': 950,
  'cover_imgurl': 'http://www.sinaimg.cn/dy/slidenews/26_img/2011_32/18496_59078_243228.jpg',
  'cover_width': 950,
  'dsptime': '',
```

```
'group_title': '徐宁作品欣赏',
'grpseq': 1,
'id': 'ed8686ac7f10dfb52d68baca348a08be',
'imageid': '51c2b804fb6d402486737c29c5301a84',
'index': 33,
'label': '',
'qhimg_height': 238,
'qhimg_thumb_url': 'http://p2.so.qhmsg.com/sdr/238__/t017f259639fd6c8287.jpg',
'qhimg_url': 'http://p2.so.qhmsg.com/t017f259639fd6c8287.jpg',
'qhimg_width': 238,
'tag': '新浪艺术名家人物库',
'total_count': 161},

...省略中间部分...

{'cover_height': 377,
'cover_imgurl': 'http://www.sinaimg.cn/dy/slidenews/26_img/2011_03/16418_23122_876413.jpg',
'cover_width': 950,
'dsptime': '',
'group_title': '王国斌作品欣赏',
'grpseq': 1,
'id': '8e173e45250d90d2dc7316777e2be59b',
'imageid': 'c7d7e74dc18685f5c100d235522d5e4b',
'index': 59,
'label': '',
'qhimg_height': 94,
'qhimg_thumb_url': 'http://p2.so.qhimgs1.com/sdr/238__/t014d248b01108afebe.jpg',
'qhimg_url': 'http://p2.so.qhimgs1.com/t014d248b01108afebe.jpg',
'qhimg_width': 238,
'tag': '新浪艺术名家人物库',
'total_count': 13},
{'cover_height': 1215,
'cover_imgurl': 'http://www.sinaimg.cn/dy/slidenews/26_img/2011_09/17732_26034_613620.jpg',
'cover_width': 900,
'dsptime': '',
'group_title': '王习三作品欣赏',
'grpseq': 1,
'id': '989031bb929f667f8eb920cfa21e32fa',
```

```
'imageid': 'f57b9882a93265edcd85e59d3fbb8a4c',
'index': 60,
'label': '王习三',
'qhimg_height': 321,
'qhimg_thumb_url': 'http://p4.so.qhmsg.com/sdr/238__/t015381735d7c0aa2a9.jpg',
'qhimg_url': 'http://p4.so.qhmsg.com/t015381735d7c0aa2a9.jpg',
'qhimg_width': 238,
'tag': '新浪艺术名家人物库',
'total_count': 31}]}
```

如上所示，响应结果（json）中的 list 字段是一个图片信息列表，count 字段是列表中图片信息的数量，每一项图片信息的 qhimg_url 字段是图片下载地址。

连续滚动鼠标滚轮加载图片，捕获更多 jQuery 发送的请求：

第 1 次加载：http://image.so.com/zj?ch=art&sn=30&listtype=new&temp=1
第 2 次加载：http://image.so.com/zj?ch=art&sn=60&listtype=new&temp=1
第 3 次加载：http://image.so.com/zj?ch=art&sn=90&listtype=new&temp=1
……

经过观察，可以总结出这些 url 的规律：

- ch 参数　分类标签。
- sn 参数　从第几张图片开始加载，即结果列表中第一张图片在服务器端的序号。

我们可以通过这个 API 每次获取固定数量的图片信息，从中提取每一张图片的 url（下载地址），直到响应结果中的 count 字段为 0（意味着没有更多图片了）。

到此，页面分析工作完成了。

9.3.3 编码实现

接下来，我们按以下 3 步完成该项目：

（1）创建 Scrapy 项目，并使用 scrapy genspider 命令创建 Spider。
（2）在配置文件中启用 ImagesPipeline，并指定图片下载目录。
（3）实现 ImagesSpider。

步骤 01 首先创建 Scrapy 项目，取名为 so_image，再使用 scrapy genspider 命令创建 Spider。

```
$ scrapy startproject so_image
$ cd so_image
$ scrapy genspider images image.so.com
```

步骤 02 在配置文件 settings.py 中启用 ImagesPipeline,并指定图片下载目录,代码如下:

```
ITEM_PIPELINES = {
    'scrapy.pipelines.images.ImagesPipeline': 1,
}
IMAGES_STORE = 'download_images'
```

步骤 03 实现 IamgesSpider,代码如下:

```
# -*- coding: utf-8 -*-
import scrapy
from scrapy import Request
import json


class ImagesSpider(scrapy.Spider):
    BASE_URL = 'http://image.so.com/zj?ch=art&sn=%s&listtype=new&temp=1'
    start_index = 0

    # 限制最大下载数量,防止磁盘用量过大
    MAX_DOWNLOAD_NUM = 1000

    name = "images"
    start_urls = [BASE_URL % 0]

    def parse(self, response):
        # 使用 json 模块解析响应结果
        infos = json.loads(response.body.decode('utf-8'))
        # 提取所有图片下载 url 到一个列表, 赋给 item 的'image_urls'字段
        yield {'image_urls': [info['qhimg_url'] for info in infos['list']]}

        # 如 count 字段大于 0,并且下载数量不足 MAX_DOWNLOAD_NUM,继续获取下一页图片信息
        self.start_index += infos['count']
        if infos['count'] > 0 and self.start_index < self.MAX_DOWNLOAD_NUM:
            yield Request(self.BASE_URL % self.start_index)
```

编码完成后,运行爬虫:

```
$ scrapy crawl images
```

运行结束后,查看图片下载目录 download_images,如图 9-7 所示,我们成功爬取到了 607 张艺术图片。

图 9-7

到此,图片下载的项目完成了。

9.4 本章小结

本章我们学习了在 Scrapy 中下载文件和图片的方法,先简单介绍了 Scrapy 提供的 FilesPipeline 和 ImagesPipeline,然后通过两个实战项目演示了 FilesPipeline 和 ImagesPipeline 的使用。

第 10 章

模拟登录

目前，大部分网站都具有用户登录功能，其中某些网站只有在用户登录后才能获得有价值的信息，在爬取这类网站时，Scrapy 爬虫程序需要先模拟用户登录，再爬取内容，这一章来学习在 Scrapy 中模拟登录的方法。

10.1 登录实质

在学习模拟登录前，应先对网站登录的原理有所了解，我们在 Chrome 浏览器中跟踪一次实际的登录操作，观察浏览器与网站服务器是如何交互的。以登录 http://example.webscraping.com 网站为例进行演示，这是一个专门用于练习爬虫技术的网站，图 10-1 所示为该网站的登录页面。

第 10 章 模拟登录

图 10-1

页面中的表单对应于 HTML 中的<form>元素,当填写完表单,单击"提交"按钮时,浏览器会根据<form>元素的内容发送一个 HTTP 请求给服务器,其中:

- <form>的 method 属性决定了 HTTP 请求的方法(本例中为 POST)。
- <form>的 action 属性决定了 HTTP 请求的 url(本例中为#,也就是当前页面的 url)。
- <form>的 enctype 属性决定了表单数据的编码类型(本例中为 x-www-urlencoded)。
- <form>中的<input>元素决定了表单数据的内容。

再来看<form>中的<input>元素:

- name 属性为'email'和'password'的两个<input>,对应于账号和密码输入框,它们的值待用户填写。
- 在<div style="display:none;">中还包含了 3 个隐藏的<input type="hidden">,它们的值在 value 属性中,虽然值不需要用户填写,但提交的表单数据中缺少它们可能会导致登录验证失败,这些隐藏的<input>有其他一些用途,比如:
  ▷ <input name="_next">用来告诉服务器,登录成功后页面跳转的地址。
  ▷ <input name="_formkey">用来防止 CSRF 跨域攻击,相关内容请查阅资料。

接下来,填入账号密码,在 Chrome 开发者工具中观察单击 Log In 按钮后浏览器发送的 HTTP 请求,如图 10-2 所示。

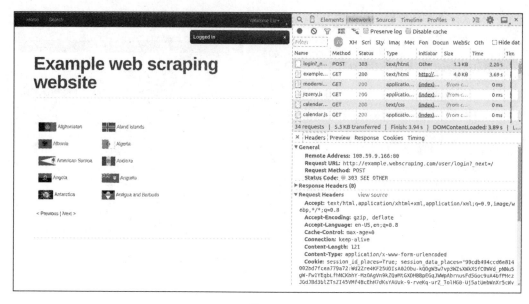

图 10-2

从图 10-2 中可以看出,我们捕获到了很多 HTTP 请求,其中第一个就是发送登录表单的 POST 请求,查看该 HTTP 请求,可以找到请求 url、请求方法(method)、请求头部(headers)、表单数据(form data)等信息,其中我们最关心的是表单数据,如图 10-3 所示。

```
▼Form Data        view source        view URL encoded
    email: liushuo@webscraping.com
    password: 12345678
    _next: /
    _formkey: 668e47e1-552d-482f-b129-64fe7b950533
    _formname: login
```

图 10-3

表单数据由多个键值对构成,每个键值对对应一个<input>元素,其中的键是<input>的 name 属性,值是<input>的 value 属性(用户填写的内容会成为<input>的 value)。需要注意的是,图 10-3 中显示的并不是实际的 POST 正文(content),这样的显示方式只是为了方便用户查看。单击 view source 按钮,可以看到实际的 POST 正文,如图 10-4 所示。

```
▼ Form Data      view parsed
    email=liushuo%40webscraping.com&password=12345678&_next=%2F&_formke
  y=668e47e1-552d-482f-b129-64fe7b950533&_formname=login
```

图 10-4

分析过 HTTP 请求信息后,再来看 HTTP 响应信息,如图 10-5 所示。

```
▼ General
    Remote Address: 108.59.9.166:80
    Request URL: http://example.webscraping.com/user/login?_next=/
    Request Method: POST
    Status Code: ● 303 SEE OTHER
▼ Response Headers      view source
    Connection: keep-alive
    Content-Length: 45
    Content-Type: text/html; charset=UTF-8
    Date: Tue, 28 Mar 2017 03:09:59 GMT
    Location: /
    Server: nginx
    Set-Cookie: session_id_places=True; Path=/
    Set-Cookie: session_data_places="c28757af06ff0ce7feac1c9cfaf54d72:2pqn
  f0ELWzT-UKYgQ8sOiW-eJs1G0sahvb5Fgbk0gaTZz5Z12pdax1lrfYJswBDlMtMKy-A2EQ
  3ohlk28fDYRK-LsEmMvo6diDy-ZiRefIIGB9J2PgKvuPdVrZqwWEMlNVUOaWFajZGijpDvi
  d52IYYXrTEbg2d2Bpt8xhLinwarWfqgd4gEfqN9yLkI12Dnwk3zTYDv5CYd63KEAGdUlzu
  JstGt-7lG3fagkehEm40XDFd0i0Nosj_QEffI49HeVcz3IIRYFIaixJcqW3mCKogyZ-Foc
  prHb4_WrSdlDgs_DGKbRoVZW3lAl9wnB6mwbRoR7cHEdwULOOWkJCSinsJRo4LLUXlleGz
  BldLsRB168hq8zPAPefXHM49yotGC_hE77ii9R3QyBWNc0WBrFfgJVM2-opsc3akoXaelB
  Birt4CJXYo68n_3xF0lpjr6HPjAE0wFrOg0ilXJdkw2MKm5vqqgO-w3sObApXngfEg2-RP
  Vk9BErzerQYxMl48PW6lvAkbwM2OUxlleFzEZ0H-liN9tej7lJcWuCvnhqnjtQjnMH93Qo
  gHQ_IhZo2RjxtLKn5dNtMsV2Yu3ePlJJ6s4yAMMvzIkB-QYztip-35OK3fcgwR17vbRct4
  ALKmJg6URAoIm2AGat4-_ahJy75lxn-b7yuMiHtetg6NQ9xxJLU1KwuoA_HytYhZGDiyTU
  8l2WuQFMzj3okNpORGijD5VA21tiY6U-G4WJtcp7t7SKP6fwES8NY0bOFzlz7mGm-c67rS
  knq7WI8kXifq1dTX0wjTc7BD2VXjvygri_MvumfdpAt8UMU13ek7iWlccqO4dIf2ciLCgo
  yGBfBJ69aFhXwsy5Rx0IOJh0ltbuXVjJQt-Z_hx1D7cBXL0jMXV1aqi5OHB7t0CIHlGHgY
  hbMlRCZmeS2HpBQr_kg4s6hlAcdB_eDbbkaYqAOFq6xgrRQ5lfsG7M7dP4XHL3YQ7bgWI1
  74dRARqz4EliMNW13_Da-KgaHNxZ60ZzOdyNkB_WakgzJXZ4-ZrZp5cBjRXP-Q=="; Pat
  h=/
```

图 10-5

响应头部中长长的 Set-Cookie 字段就是网站服务器程序保存在客户端(浏览器)的 Cookie 信息,其中包含标识用户身份的 session 信息,之后对该网站发送的其他 HTTP 请求都会带上这个"身份证"(session 信息),服务器程序通过这个"身份证"识别出发送请求的用户,从而决定响应怎样的页面。另外,响应的状态码是 303,它代表页面重定向,浏览器会读取响应头部中的 Location 字段,依据其中描述的路径(本例中为/)再次发送一个 GET 请求,图 10-6 所示为这个 GET 请求的信息。

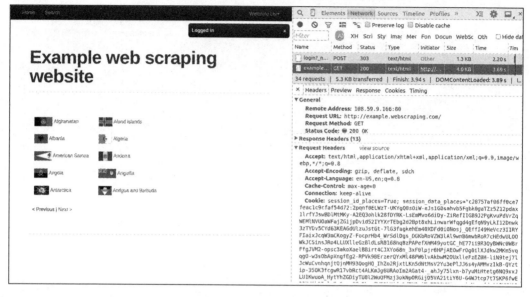

图 10-6

观察请求头部中的 Cookie 字段,它携带了之前 POST 请求获取的 Cookie 信息,最终浏览器用该请求响应的 HTML 文档刷新了页面,在页面右上角可以看到 Welcome Liu 的字样,表明这是登录成功后的页面。

10.2 Scrapy 模拟登录

10.2.1 使用 FormRequest

现在大家了解了登录的实质,其核心是向服务器发送含有登录表单数据的 HTTP 请求(通常是 POST)。Scrapy 提供了一个 FormRequest 类(Request 的子类),专门用于构造含有表单数据的请求,FormRequest 的构造器方法有一个 formdata 参数,接收字典形式的表单数据。接下来,我们在 scrapy shell 环境下演示如何使用 FormRequest 模拟登录。

首先爬取登录页面 http://example.webscraping.com/user/login:

```
$ scrapy shell http://example.webscraping.com/user/login
```

通过之前的分析,我们已经了解了表单数据应包含的信息:账号和密码信息,再加3个隐藏<input>中的信息。先把这些信息收集到一个字典中,然后使用这个表单数据字典构造 FormRequest 对象:

```
>>> # 先提取 3 个隐藏<input>中包含的信息,它们在<div style="display:none;">中
>>> sel = response.xpath('//div[@style]/input')
>>> sel
[<Selector xpath='//div[@style]/input' data='<input name="_next" type="hidden" value='>,
 <Selector xpath='//div[@style]/input' data='<input name="_formkey" type="hidden" val'>,
 <Selector xpath='//div[@style]/input' data='<input name="_formname" type="hidden" va'>]
>>> # 构造表单数据字典
>>> fd = dict(zip(sel.xpath('./@name').extract(), sel.xpath('./@value').extract()))
>>> fd
{'_formkey': '432dcb0c-0d85-443f-bb50-9644cfeb2f2b',
 '_formname': 'login',
 '_next': '/'}
>>> # 填写账号和密码信息
>>> fd['email'] = 'liushuo@webscraping.com'
>>> fd['password'] = '12345678'
>>> fd
{'_formkey': '432dcb0c-0d85-443f-bb50-9644cfeb2f2b',
 '_formname': 'login',
 '_next': '/',
 'email': 'liushuo@webscraping.com',
 'password': '12345678'}
>>> from scrapy.http import FormRequest
>>> request = FormRequest('http://example.webscraping.com/user/login', formdata=fd)
```

以上是直接构造 FormRequest 对象的方式,除此之外还有一种更为简单的方式,即调用 FormRequest 的 from_response 方法。调用时需传入一个 Response 对象作为第一个参数,该方法会解析 Response 对象所包含页面中的<form>元素,帮助用户创建 FormRequest 对象,并将隐藏<input>中的信息自动填入表单数据。使用这种方式,我们只需通过 formdata 参数填写账号和密码即可,代码如下:

```
>>> fd = {'email': 'liushuo@webscraping.com', 'password': '12345678'}
>>> request = FormRequest.from_response(response, formdata=fd)
```

使用任意方式构造好 FormRequest 对象后，接下来提交表单请求：

```
>>> fetch(request)
[scrapy] DEBUG: Redirecting (303) to <GET http://example.webscraping.com/> from
<POST http://example.webscraping.com/user/login>
[scrapy] DEBUG: Crawled (200) <GET http://example.webscraping.com/> (referer: None)
```

在 log 信息中，可以看到和浏览器提交表单时类似的情形：POST 请求的响应状态码为 303，之后 Scrapy 自动再发送一个 GET 请求下载跳转页面。此时，可以通过在页面中查找特殊字符串或在浏览器中查看页面验证登录是否成功，如图 10-7 所示。

```
>>> 'Welcome Liu' in response.text
True
>>> view(response)
```

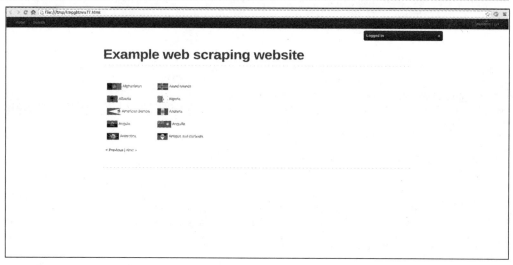

图 10-7

验证结果表明模拟登录成功了。显然，Scrapy 发送的第 2 个 GET 请求携带了第 1 个 POST 请求获取的 Cookie 信息，为请求附加 Cookie 信息的工作是由 Scrapy 内置的下载中间件 CookiesMiddleware 自动完成的。现在，我们可以继续发送请求，爬取那些只有登录后才能获取的信息了，这里以爬取用户个人信息为例，如图 10-8 所示。

```
>>> fetch('http://example.webscraping.com/user/profile')    #下载用户个人信息页面
[scrapy] DEBUG: Crawled (200) <GET http://example.webscraping.com/user/profile>
>>> view(response)
```

图 10-8

提取页面中的用户个人信息，代码如下：

```
>>> keys = response.css('table label::text').re('(.+):')
>>> keys
['First name', 'Last name', 'E-mail']
>>> values = response.css('table td.w2p_fw::text').extract()
>>> values
['Liu', 'Shuo', 'liushuo@webscraping.com']
>>> dict(zip(keys, values))
{'E-mail': 'liushuo@webscraping.com', 'First name': 'Liu', 'Last name': 'Shuo'}
```

到此，使用 FormRequest 模拟登录的过程就演示完了。

10.2.2 实现登录 Spider

整理 10.2.1 小节中登录 http://example.webscraping.com 后爬取用户个人信息的代码，实现一个 LoginSpider，代码如下：

```
# -*- coding: utf-8 -*-
import scrapy
from scrapy.http import Request, FormRequest
```

```python
class LoginSpider(scrapy.Spider):
    name = "login"
    allowed_domains = ["example.webscraping.com"]
    start_urls = ['http://example.webscraping.com/user/profile']

    def parse(self, response):
        # 解析登录后下载的页面，此例中为用户个人信息页面
        keys = response.css('table label::text').re('(.+):')
        values = response.css('table td.w2p_fw::text').extract()

        yield dict(zip(keys, values))

    # ---------------------------登录--------------------------------
    # 登录页面的 url
    login_url = 'http://example.webscraping.com/user/login'

    def start_requests(self):
        yield Request(self.login_url, callback=self.login)

    def login(self, response):
        # 登录页面的解析函数，构造 FormRequest 对象提交表单
        fd = {'email': 'liushuo@webscraping.com', 'password': '12345678'}
        yield FormRequest.from_response(response, formdata=fd,
                                        callback=self.parse_login)

    def parse_login(self, response):
        # 登录成功后，继续爬取 start_urls 中的页面
        if 'Welcome Liu' in response.text:
            yield from super().start_requests()          # Python 3 语法
```

解释上述代码如下：

- 覆写基类的 start_requests 方法，最先请求登录页面。
- login 方法为登录页面的解析函数，在该方法中进行模拟登录，构造表单请求并提交。
- parse_login 方法为表单请求的响应处理函数，在该方法中通过在页面查找特殊字符串'Welcome Liu'判断是否登录成功，如果成功，调用基类的 start_requests 方法，继续爬取 start_urls 中的页面。

我们这样设计 LoginSpider 就是想把模拟登录和爬取内容的代码分离开，使得逻辑上更加清晰。

10.3　识别验证码

目前，很多网站为了防止爬虫爬取，登录时需要用户输入验证码。下面我们学习如何在爬虫程序中识别验证码（在举例过程中不指明具体网站）。

图 10-9 所示为某网站登录页面，其中包含验证码。

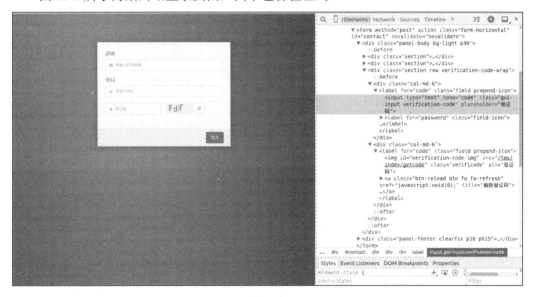

图 10-9

页面中的验证码图片对应一个<img>元素，即一张图片，浏览器加载完登录页面后，会携带之前访问获取的 Cookie 信息，继续发送一个 HTTP 请求加载验证码图片。和账号密码输入框一样，验证码输入框也对应一个<input>元素，因此用户输入的验证码会成为表单数据的一部分，表单提交后由网站服务器程序验证。

识别验证码有多种方式，下面介绍常用的几种。

10.3.1　OCR 识别

OCR 是光学字符识别的缩写，用于在图像中提取文本信息，tesseract-ocr 是利用该技术实现的一个验证码识别库，在 Python 中可以通过第三方库 pytesseract 调用它。下面介绍如何使用 pytesseract 识别验证码。

首先安装 tesseract-ocr,在 Ubuntu 下可以使用 apt-get 安装:

```
sudo apt-get install tesseract-ocr
```

接下来安装 pytesseract,它依赖于 Python 图像处理库 PIL 或 Pillow(PIL 和 Pillow 功能类似,任选其一),可以使用 pip 安装它们:

```
$ pip install pillow
$ pip install pytesseract
```

现在,我们使用 pytesseract 识别图片 code.png 中的验证码,代码如下:

```
>>> from PIL import Image
>>> import pytesseract
>>> img = Image.open('code.png')
>>> img = img.convert('L')
>>> pytesseract.image_to_string(img)
'cqKE'
```

上面的代码中,先使用 Image.open 打开图片,为了提高识别率,调用 Image 对象的 convert 方法把图片转换为黑白图,然后将黑白图传递给 pytesseract.image_to_string 方法进行识别,这里我们幸运地识别成功了。经测试,此段代码对于 X 网站中的验证码识别率可以达到 72%,这已经足够高了。

下面我们以之前的 LoginSpider 为模板实现一个使用 pytesseract 识别验证码登录的 Spider:

```
# -*- coding: utf-8 -*-
import scrapy
from scrapy import Request, FormRequest
import json
from PIL import Image
from io import BytesIO
import pytesseract
from scrapy.log import logger

class CaptchaLoginSpider(scrapy.Spider):
```

```python
name = "login_captcha"
start_urls = ['http://XXX.com/']

def parse(self, response):
    ...
# X 网站登录页面的 url（虚构的）
login_url = 'http://XXX.com/login'
user      = 'liushuo@XXX.com'
password  = '12345678'

def start_requests(self):
    yield Request(self.login_url, callback=self.login, dont_filter=True)

def login(self, response):
    # 该方法既是登录页面的解析函数，又是下载验证码图片的响应处理函数

    # 如果 response.meta['login_response']存在，当前 response 为验证码图片的响应
    # 否则当前 response 为登录页面的响应
    login_response = response.meta.get('login_response')

    if not login_response:
        # Step 1:
        # 此时 response 为登录页面的响应，从中提取验证码图片的 url，下载验证码图片
        captchaUrl = response.css('label.field.prepend-icon img::attr(src)')\
                        .extract_first()

        captchaUrl = response.urljoin(captchaUrl)
        # 构造 Request 时，将当前 response 保存到 meta 字典中
        yield Request(captchaUrl,
                callback=self.login,
                meta={'login_response': response},
                dont_filter=True)

    else:
        # Step 2:
        # 此时，response 为验证码图片的响应，response.body 是图片二进制数据
        # login_response 为登录页面的响应，用其构造表单请求并发送
        formdata = {
            'email': self.user,
            'pass': self.password,
            # 使用 OCR 识别
            'code': self.get_captcha_by_OCR(response.body),
```

```
            }
            yield FormRequest.from_response(login_response,
                                            callback=self.parse_login,
                                            formdata=formdata, dont_filter=True)

def parse_login(self, response):
    # 根据响应结果判断是否登录成功
    info = json.loads(response.text)
    if info['error'] == '0':
        logger.info('登录成功:-)')
        return super().start_requests()

    logger.info('登录失败:-(, 重新登录...')
    return self.start_requests()

def get_captcha_by_OCR(self, data):
    # OCR 识别
    img = Image.open(BytesIO(data))
    img = img.convert('L')
    captcha = pytesseract.image_to_string(img)
    img.close()

    return captcha
```

解释上述代码如下:

- login 方法

  带有验证码的登录,需要额外发送一个 HTTP 请求来获取验证码图片,这里的 login 方法既处理下载登录页面的响应,又处理下载验证码图片的响应。

  ➤ 解析登录页面时,提取验证码图片的 url,发送请求下载图片,并将登录页面的 Response 对象保存到 Request 对象的 meta 字典中。

  ➤ 处理下载验证码图片的响应时,调用 get_captcha_by_OCR 方法识别图片中的验证码,然后将之前保存的登录页面的 Response 对象取出,构造 FormRequest 对象并提交。

- get_captcha_by_OCR 方法

  参数 data 是验证码图片的二进制数据,类型为 bytes,想要使用 Image.open 函数构造 Image 对象,先要把图片的二进制数据转换成某种类文件对象,这里使用 BytesIO 进行包裹,获得 Image 对象后先将其转换成黑白图,然后调用 pytesseract.image_to_string 方法进行识别。

- parse_login 方法

  处理表单请求的响应。响应正文是一个 json 串,其中包含了用户验证的结果,先调用 json.loads 将正文转换为 Python 字典,然后依据其中 error 字段的值判断登录是否成功,若登录成功,则从 start_urls 中的页面开始爬取;若登录失败,则重新登录。

10.3.2 网络平台识别

在 10.3.1 小节中,使用 pytesseract 识别的验证码比较简单,对于某些复杂的验证码,pytesseract 的识别率很低或者无法识别。目前,有很多网站专门提供验证码识别服务,可以识别较为复杂的验证码(有些是人工处理的),它们被称之为验证码识别平台,这些平台多数是付费使用的,价格大约为 1 元钱识别 100 个验证码,平台提供了 HTTP 服务接口,用户可以通过 HTTP 请求将验证码图片发送给平台,平台识别后将结果通过 HTTP 响应返回。

在阿里云市场可以找到很多验证码识别平台,我们随意挑选了一个,如图 10-10 和图 10-11 所示。

图 10-10

图片验证码识别_中英数

调用地址：http://ali-checkcode.showapi.com/checkcode

请求方式：POST

返回类型：JSON

API 调用：API 简单身份认证调用方法（APPCODE）展开▼

调试工具： 去调试

▶ 请求参数（Headers）

▶ 请求参数（Query）

▼ 请求参数（Body）

名称	类型	是否必须	描述
convert_to_jpg	STRING	可选	有少量png或gif图转成jpg格式后识别率明显提高（并不是所有png或gif转成jpg后都会提高识别率）。此字段为1时表示需要把图片转为jpg格式，其他值不做转换操作。
img_base64	STRING	必选	图片文件的base64字符串。图片大小需要小于100KB。
typeId	STRING	必选	1.纯数字 typeId=1000 任意长度数字，识别率会降低 typeId=1010 1位数字 typeId=1020 2位数字 ... typeId=1100 10位数字 2.纯英文 typeId=2000 任意长度英文，识别率会降低 typeId=2010 1位英文 typeId=2020 2位英文 ... typeId=2100 10位英文 3.英文数字混合 typeId=3000 任意长度英数混合，识别率会降低 typeId=3010 1位英数 typeId=3020 2位英数混合 ... typeId=3100 10位英数混合 4.纯汉字 typeId=4000 任意长度汉字混合，识别率会降低 typeId=4010 1位汉字 typeId=4020 2位汉字 ... typeId=4100 10位

图 10-11

购买服务后，我们利用该平台识别图片 code.gif 中较为复杂的验证码：

阅读 API 文档，实现代码如下：

```
import requests
import base64
from pprint import pprint

# 购买服务后，平台发放给我们一个 appcode，用来识别请求者的身份
APPCODE = 'f239ccawf37f287418a90e2f922649273c4'

url = 'http://ali-checkcode.showapi.com/checkcode'
img_data = open('code.gif', 'rb').read()

form = {}

# 不转换为 jpg
form['convert_to_jpg'] = '0'
# 对图片进行 base64 编码
form['img_base64'] = base64.b64encode(img_data)
# 7 位汉字
form['typeId'] = '4070'

# 用户验证
headers = {'Authorization': 'APPCODE ' + APPCODE}
response = requests.post(url, headers=headers, data=form)
pprint(response.json())
```

表单中 3 个字段含义如下：

- convert_to_jpg　是否将图片转换为 jpg 格式。
- img_base64　图片数据的 base64 编码。
- typeId　验证码类型，这里的 '4070' 代表 7 位汉字。

运行脚本,并观察结果:

```
$ python3 ali_checkcode.py
{'showapi_res_body': {'Id': '4d5fea-21eb-4043-8236-d478031916',
                      'Result': '损俱饶现渊弹翠',
                      'ret_code': 0},
 'showapi_res_code': 0,
 'showapi_res_error': ''}
```

如上所示,识别成功了。

现在,我们把使用平台识别验证码的方式也加入 CaptchaLoginSpider 中,只需添加一个 get_captcha_by_network 方法:

```python
class CaptchaLoginSpider(scrapy.Spider):
    ...

    def get_captcha_by_OCR(self, data):
        # OCR 识别
        img = Image.open(BytesIO(data))
        img = img.convert('L')
        captcha = pytesseract.image_to_string(img)
        img.close()

        return captcha

    def get_captcha_by_network(self, data):
        # 平台识别
        import requests

        url = 'http://ali-checkcode.showapi.com/checkcode'
        appcode = 'f23cca37f287418a90e2f922649273c4'

        form = {}
        form['convert_to_jpg'] = '0'
        form['img_base64'] = base64.b64encode(data)
        form['typeId'] = '3040'

        headers = {'Authorization': 'APPCODE ' + appcode}
        response = requests.post(url, headers=headers, data=form)
```

```
        res = response.json()

        if res['showapi_res_code'] == 0:
            return res['showapi_res_body']['Result']

        return ''
```

10.3.3 人工识别

最后讲解的方法听起来似乎很笨：人工识别。通常网站只需登录一次便可爬取，在其他识别方式不管用时，人工识别一次验证码也是可行的，其实现也非常简单——在 Scrapy 下载完验证码图片后，调用 Image.show 方法将图片显示出来，然后调用 Python 内置的 input 函数，等待用户肉眼识别后输入识别结果。

我们将人工识别的方式也加入 CaptchaLoginSpider 中，再添加一个 get_captcha_by_user 方法：

```
class CaptchaLoginSpider(scrapy.Spider):
    ...

    def get_captcha_by_OCR(self, data):
        # OCR 识别
        img = Image.open(BytesIO(data))
        img = img.convert('L')
        captcha = pytesseract.image_to_string(img)
        img.close()

        return captcha

    def get_captcha_by_network(self, data):
        # 平台识别
        import requests

        url = 'http://ali-checkcode.showapi.com/checkcode'
        appcode = 'f23cca37f287418a90e2f922649273c4'

        form = {}
        form['convert_to_jpg'] = '0'
        form['img_base64'] = base64.b64encode(data)
```

```python
        form['typeId'] = '3040'

        headers = {'Authorization': 'APPCODE ' + appcode}
        response = requests.post(url, headers=headers, data=form)
        res = response.json()

        if res['showapi_res_code'] == 0:
            return res['showapi_res_body']['Result']

        return ''

    def get_captcha_by_user(self, data):
        # 人工识别
        img = Image.open(BytesIO(data))
        img.show()
        captcha = input('输入验证码:')
        img.close()
        return captcha
```

10.4　Cookie 登录

在 10.3 节，我们讲解了识别验证码登录的方法，但目前网站的验证码越来越复杂，某些验证码已经复杂到人类难以识别的程度，有些时候提交表单登录的路子难以走通。此时，我们可以换一种登录爬取的思路，在使用浏览器登录网站后，包含用户身份信息的 Cookie 会被浏览器保存在本地，如果 Scrapy 爬虫能直接使用浏览器中的 Cookie 发送 HTTP 请求，就可以绕过提交表单登录的步骤。

10.4.1　获取浏览器 Cookie

我们无须费心钻研，各种浏览器将 Cookie 以哪种形式存储在哪里，使用第三方 Python 库 browsercookie 便可以获取 Chrome 和 Firefox 浏览器中的 Cookie。

使用 pip 安装 browsercookie：

```
pip install browsercookie
```

browsercookie 的使用非常简单，示例代码如下：

```
>>> import browsercookie
>>> chrome_cookiejar = browsercookie.chrome()      # 获取 Chrome 浏览器中的 Cookie
>>> firefox_cookiejar = browsercookie.firefox()    # 获取 Firefox 浏览器中的 Cookie
>>> type(chrome_cookiejar)
http.cookiejar.CookieJar
>>> for cookie in chrome_cookiejar:
...     print(cookie)
```

browsercookie 的 chrome 和 firefox 方法分别返回 Chrome 和 Firefox 浏览器中的 Cookie，返回值是一个 http.cookiejar.CookieJar 对象，对 CookieJar 对象进行迭代，可以访问其中的每个 Cookie 对象。

10.4.2 CookiesMiddleware 源码分析

之前曾提到过，Scrapy 爬虫所使用的 Cookie 由内置下载中间件 CookiesMiddleware 自动处理。下面我们来分析一下 CookiesMiddleware 是如何工作的，其源码如下：

```python
import os
import six
import logging
from collections import defaultdict

from scrapy.exceptions import NotConfigured
from scrapy.http import Response
from scrapy.http.cookies import CookieJar
from scrapy.utils.python import to_native_str

logger = logging.getLogger(__name__)

class CookiesMiddleware(object):
    """This middleware enables working with sites that need cookies"""

    def __init__(self, debug=False):
        self.jars = defaultdict(CookieJar)
        self.debug = debug

    @classmethod
    def from_crawler(cls, crawler):
        if not crawler.settings.getbool('COOKIES_ENABLED'):
```

```
            raise NotConfigured
        return cls(crawler.settings.getbool('COOKIES_DEBUG'))

    def process_request(self, request, spider):
        if request.meta.get('dont_merge_cookies', False):
            return

        cookiejarkey = request.meta.get("cookiejar")
        jar = self.jars[cookiejarkey]
        cookies = self._get_request_cookies(jar, request)
        for cookie in cookies:
            jar.set_cookie_if_ok(cookie, request)

        # set Cookie header
        request.headers.pop('Cookie', None)
        jar.add_cookie_header(request)
        self._debug_cookie(request, spider)

    def process_response(self, request, response, spider):
        if request.meta.get('dont_merge_cookies', False):
            return response

        # extract cookies from Set-Cookie and drop invalid/expired cookies
        cookiejarkey = request.meta.get("cookiejar")
        jar = self.jars[cookiejarkey]
        jar.extract_cookies(response, request)
        self._debug_set_cookie(response, spider)

        return response

    def _debug_cookie(self, request, spider):
        if self.debug:
            cl = [to_native_str(c, errors='replace')
                  for c in request.headers.getlist('Cookie')]
            if cl:
                cookies = "\n".join("Cookie: {}\n".format(c) for c in cl)
                msg = "Sending cookies to: {}\n{}".format(request, cookies)
                logger.debug(msg, extra={'spider': spider})
```

```python
def _debug_set_cookie(self, response, spider):
    if self.debug:
        cl = [to_native_str(c, errors='replace')
              for c in response.headers.getlist('Set-Cookie')]
        if cl:
            cookies = "\n".join("Set-Cookie: {}\n".format(c) for c in cl)
            msg = "Received cookies from: {}\n{}".format(response, cookies)
            logger.debug(msg, extra={'spider': spider})

def _format_cookie(self, cookie):
    # build cookie string
    cookie_str = '%s=%s' % (cookie['name'], cookie['value'])

    if cookie.get('path', None):
        cookie_str += '; Path=%s' % cookie['path']
    if cookie.get('domain', None):
        cookie_str += '; Domain=%s' % cookie['domain']

    return cookie_str

def _get_request_cookies(self, jar, request):
    if isinstance(request.cookies, dict):
        cookie_list = [{'name': k, 'value': v} for k, v in \
                       six.iteritems(request.cookies)]
    else:
        cookie_list = request.cookies

    cookies = [self._format_cookie(x) for x in cookie_list]
    headers = {'Set-Cookie': cookies}
    response = Response(request.url, headers=headers)

    return jar.make_cookies(response, request)
```

分析其中几个核心方法如下：

- **from_crawler 方法**
  从配置文件中读取 COOKIES_ENABLED，决定是否启用该中间件。如果启用，调用构造器创建对象，否则抛出 NotConfigured 异常，Scrapy 将忽略该中间件。

- __init__ 方法

  使用标准库中的 collections.defaultdict 创建一个默认字典 self.jars，该字典中每一项的值都是一个 scrapy.http.cookies.CookieJar 对象，CookiesMiddleware 可以让 Scrapy 爬虫同时使用多个不同的 CookieJar。例如，在某网站我们注册了两个账号 account1 和 account2，假设一个爬虫想同时登录两个账号对网站进行爬取，为了避免 Cookie 冲突（通俗地讲，登录一个会替换掉另一个），此时可以让每个账号发送的 HTTP 请求使用不同的 CookieJar，在构造 Request 对象时，可以通过 meta 参数的 cookiejar 字段指定所要使用的 CookieJar，如：

```
# 账号 account1 发送的请求
Request(url1, meta={'cookiejar': 'account1'})
Request(url2, meta={'cookiejar': 'account1'})
Request(url3, meta={'cookiejar': 'account1'})
...
# 账号 account2 发送的请求
Request(url1, meta={'cookiejar': 'account2'})
Request(url2, meta={'cookiejar': 'account2'})
Request(url3, meta={'cookiejar': 'account2'})
...
```

- process_request 方法

  处理每一个待发送的 Request 对象，尝试从 request.meta['cookiejar'] 获取用户指定使用的 CookieJar，如果用户未指定，就使用默认的 CookieJar(self.jars[None])。调用 self._get_request_cookies 方法获取发送请求 request 应携带的 Cookie 信息，填写到 HTTP 请求头部。

- process_response 方法

  处理每一个 Response 对象，依然通过 request.meta['cookiejar'] 获取 CookieJar 对象，调用 extract_cookies 方法将 HTTP 响应头部中的 Cookie 信息保存到 CookieJar 对象中。

另外需要注意的是，这里的 CookieJar 是 scrapy.http.cookies.CookieJar，而 10.4.1 小节中的 CookieJar 是标准库中的 http.cookiejar.CookieJar，它们是不同的类，前者对后者进行了包装，两者可以相互转化。

10.4.3 实现 BrowserCookiesMiddleware

CookiesMiddleware 自动处理 Cookie 的特性给用户提供了便利，但它不能使用浏览器的 Cookie，我们可以利用 browsercookie 对 CookiesMiddleware 进行改良，实现一个能使用浏览器 Cookie 的中间件，代码如下：

```python
import browsercookie
from scrapy.downloadermiddlewares.cookies import CookiesMiddleware

class BrowserCookiesMiddleware(CookiesMiddleware):
    def __init__(self, debug=False):
        super().__init__(debug)
        self.load_browser_cookies()

    def load_browser_cookies(self):
        # 加载 Chrome 浏览器中的 Cookie
        jar = self.jars['chrome']
        chrome_cookiejar = browsercookie.chrome()
        for cookie in chrome_cookiejar:
            jar.set_cookie(cookie)

        # 加载 Firefox 浏览器中的 Cookie
        jar = self.jars['firefox']
        firefox_cookiejar = browsercookie.firefox()
        for cookie in firefox_cookiejar:
            jar.set_cookie(cookie)
```

了解了 CookiesMiddleware 的工作原理，便不难理解 BrowserCookiesMiddleware 的实现了，其核心思想是：在构造 BrowserCookiesMiddleware 对象时，使用 browsercookie 将浏览器中的 Cookie 提取，存储到 CookieJar 字典 self.jars 中，解释代码如下：

- 继承 CookiesMiddleware 并实现构造器方法，在构造器方法中先调用基类的构造器方法，然后调用 self.load_browser_cookies 方法加载浏览器 Cookie。
- 在 load_browser_cookies 方法中，使用 self.jars['chrome']和 self.jars['firefox']从默认字典中获得两个 CookieJar 对象，然后调用 browsercookie 的 chrome 和 firefox 方法，分别获取两个浏览器中的 Cookie，将它们填入各自的 CookieJar 对象中。

10.4.4 爬取知乎个人信息

下面通过一个例子展示 BrowserCookiesMiddleware 的使用，知乎网是国内最流行的问答网站，我们在 Chrome 浏览器登录知乎后，可以访问如图 10-12 所示的用户个人信息页面。

接下来，我们使用 BrowserCookiesMiddleware 爬取这个登录后才能访问的页面，提取用户的"姓名"和"个性域名"信息。

图 10-12

首先创建 Scrapy 项目，取名为 browser_cookie：

```
$ scrapy startproject browser_cookie
```

然后，将 BrowserCookiesMiddleware 源码复制到该项目下的 middlewares.py 中，并在配置文件 settings.py 中添加如下配置：

```
# 伪装成常规浏览器
USER_AGENT = 'Mozilla/5.0 (X11; Linux x86_64) Chrome/42.0.2311.90 Safari/537.36'
# 用 BrowserCookiesMiddleware 替代 CookiesMiddleware 启用前者，关闭后者
DOWNLOADER_MIDDLEWARES = {
    'scrapy.downloadermiddlewares.cookies.CookiesMiddleware': None,
    'browser_cookie.middlewares.BrowserCookiesMiddleware': 701,
}
```

由于需求非常简单，因此不再编写 Spider，直接在 scrapy shell 环境中进行演示。注意，为了使用项目中的配置，需要在项目目录下启动 scrapy shell 命令：

```
$ scrapy shell
...
>>> from scrapy import Request
>>> url = 'https://www.zhihu.com/settings/profile'
>>> fetch(Request(url, meta={'cookiejar': 'chrome'}))
...
>>> view(response)
```

调用 view 函数后，在浏览器中可看到如图 10-13 所示的页面。

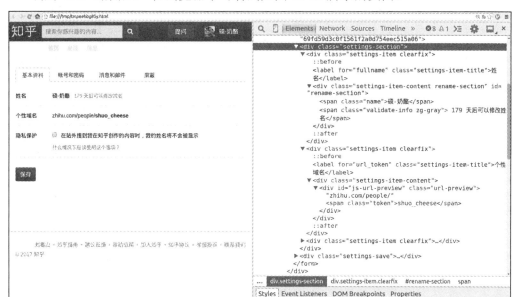

图 10-13

结果表明，BrowserCookiesMiddleware 按我们所预期的工作了，Scrapy 爬虫使用浏览器的 Cookie 成功地获取了一个需要用户登录后才能访问的页面。

最后，提取页面中的"姓名"和"个性域名"信息：

>>> response.css('div#rename-section span.name::text').extract_first()
'硕-奶酪'
>>> response.xpath('string(//div[@id="js-url-preview"])').extract_first()
'zhihu.com/people/shuo_cheese'

到此，使用浏览器 Cookie 登录的案例展示完毕。

10.5 本章小结

本章我们学习了 Scrapy 爬虫模拟登录网站的相关内容，首先介绍了网站登录的原理，并讲解如何使用 FormRequest 提交登录表单模拟登录，然后讲解识别验证码的 3 种方法，最后介绍如何使用浏览器 Cookie 直接登录，并实现了一个下载中间件 BrowserCookiesMiddleware。

第 11 章

爬取动态页面

在之前章节中，我们爬取的都是静态页面中的信息，静态页面的内容始终不变，爬取相对容易，但在现实中，目前绝大多数网站的页面都是动态页面，动态页面中的部分内容是浏览器运行页面中的 JavaScript 脚本动态生成的，爬取相对困难，这一章来学习如何爬取动态页面。

先来看一个简单的动态页面的例子，在浏览器中打开 http://quotes.toscrape.com/js，显示如图 11-1 所示。

页面中有 10 条名人名言，每一条都包含在一个<div class="quote">元素中，如图 11-2 所示。现在，我们在 scrapy shell 环境下尝试爬取页面中的名人名言：

```
$ scrapy shell http://quotes.toscrape.com/js/
...
>>> response.css('div.quote')
[]
```

从结果看出，爬取失败了，在页面中没有找到任何包含名人名言的<div class="quote">元素。这些<div class="quote">就是动态内容，从服务器下载的页面中并不包含它们（所以我们爬取失败），浏览器执行了页面中的一段 JavaScript 代码后，它们才被生成出来，如图 11-3 所示。

第 11 章 爬取动态页面　　137

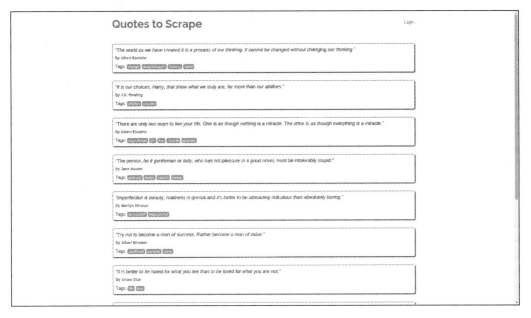

图 11-1

图 11-2

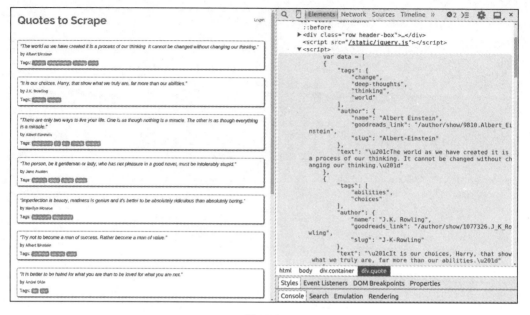

图 11-3

图中的 JavaScript 代码如下：

```
var data = [
{
    "tags": [
        "change",
        "deep-thoughts",
        "thinking",
        "world"
    ],
    "author": {
        "name": "Albert Einstein",
        "goodreads_link": "/author/show/9810.Albert_Einstein",
        "slug": "Albert-Einstein"
    },
    "text": "\u201cThe world as we have created it is a process of our thinking. \
            It cannot be changed without changing our thinking.\u201d"
},
{
    "tags": [
        "abilities",
```

```
                "choices"
            ],
            "author": {
                "name": "J.K. Rowling",
                "goodreads_link": "/author/show/1077326.J_K_Rowling",
                "slug": "J-K-Rowling"
            },
            "text": "\u201cIt is our choices, Harry, that show what we truly are, \
                far more than our abilities.\u201d"
        },
        ... 省略部分内容 ...
        {
            "tags": [
                "humor",
                "obvious",
                "simile"
            ],
            "author": {
                "name": "Steve Martin",
                "goodreads_link": "/author/show/7103.Steve_Martin",
                "slug": "Steve-Martin"
            },
            "text": "\u201cA day without sunshine is like, you know, night.\u201d"
        }
    ];
    for (var i in data) {
        var d = data[i];
        var tags = $.map(d['tags'], function(t) {
            return "<a class='tag'>" + t + "</a>";
        }).join(" ");
        document.write("<div class='quote'><span class='text'>" +
            d['text'] + "</span><span>by <small class='author'>" +
            d['author']['name'] + "</small></span><div class='tags'>Tags: " +
            tags + "</div></div>");
    }
```

　　阅读代码可以了解页面动态生成的细节，所有名人名言信息被保存在数组 data 中，最后的 for 循环迭代 data 中的每项信息，使用 document.write 生成每条名人名言对应的 <div class="quote"> 元素。

上面是动态网页中最简单的一个例子,数据被硬编码于 JavaScript 代码中,实际中更常见的是 JavaScript 通过 HTTP 请求跟网站动态交互获取数据(AJAX),然后使用数据更新 HTML 页面。爬取此类动态网页需要先执行页面中的 JavaScript 代码渲染页面,再进行爬取。下面我们介绍如何使用 JavaScript 渲染引擎渲染页面。

11.1 Splash 渲染引擎

Splash 是 Scrapy 官方推荐的 JavaScript 渲染引擎,它是使用 Webkit 开发的轻量级无界面浏览器,提供基于 HTTP 接口的 JavaScript 渲染服务,支持以下功能:

- 为用户返回经过渲染的 HTML 页面或页面截图。
- 并发渲染多个页面。
- 关闭图片加载,加速渲染。
- 在页面中执行用户自定义的 JavaScript 代码。
- 执行用户自定义的渲染脚本(lua),功能类似于 PhantomJS。

首先安装 Splash,在 linux 下使用 docker 安装十分方便:

```
$ sudo apt-get install docker
$ sudo docker pull scrapinghub/splash
```

安装完成后,在本机的 8050 和 8051 端口开启 Splash 服务:

```
$ sudo docker run -p 8050:8050 -p 8051:8051 scrapinghub/splash
[-] Log opened.
[-] Splash version: 2.1
[-] Qt 5.5.1, PyQt 5.5.1, WebKit 538.1, sip 4.17, Twisted 16.1.1, Lua 5.2
[-] Python 3.5 (default, Oct 14 2015, 20:28:29) [GCC 4.8.4]
[-] Open files limit: 524288
[-] Open files limit increased from 524288 to 1048576
[-] Xvfb is started: ['Xvfb', ':1', '-screen', '0', '1024x768x24']
[-] proxy profiles support is enabled, proxy profiles path: /etc/splash/proxy-profiles
[-] verbosity=1
[-] slots=50
[-] argument_cache_max_entries=500
[-] Web UI: enabled, Lua: enabled (sandbox: enabled)
[-] Site starting on 8050
[-] Starting factory
```

Splash 功能丰富，包含多个服务端点，由于篇幅有限，这里只介绍其中两个最常用的端点：

- render.html
  提供 JavaScript 页面渲染服务。
- execute
  执行用户自定义的渲染脚本（lua），利用该端点可在页面中执行 JavaScript 代码。

Splash 文档地址：http://splash.readthedocs.io/en/latest/api.html。

11.1.1　render.html 端点

JavaScript 页面渲染服务是 Splash 中最基础的服务，请看表 11-1 中列出的文档。

表 11-1

服务端点	render.html
请求地址	http://localhost:8050/render.html
请求方式	GET / POST
返回类型	html

render.html 端点支持的参数如表 11-2 所示。

表 11-2

参数	是否必选	类型	描述
url	必选	string	需要渲染页面的 url
timeout	可选	float	渲染页面超时时间
proxy	可选	string	代理服务器地址
wait	可选	float	等待页面渲染的时间
images	可选	integer	是否下载图片，默认为 1
js_source	可选	string	用户自定义的 JavaScript 代码，在页面渲染前执行

这里仅列出部分常用参数，详细内容参见官方文档。

下面是使用 requests 库调用 render.html 端点服务对页面 http://quotes.toscrape.com/js/ 进行渲染的示例代码。

```
>>> import requests
>>> from scrapy.selector import Selector
>>> splash_url = 'http://localhost:8050/render.html'
```

```
>>> args = {'url': 'http://quotes.toscrape.com/js', 'timeout': 5, 'image': 0}
>>> response = requests.get(splash_url, params=args)
>>> sel = Selector(response)
>>> sel.css('div.quote span.text::text').extract()     #提取所有名人名言
['"The world as we have created it is a process of our thinking. It cannot be changed without changing our thinking."',
 '"It is our choices, Harry, that show what we truly are, far more than our abilities."',
 '"There are only two ways to live your life. One is as though nothing is a miracle. The other is as though everything is a miracle."',
 '"The person, be it gentleman or lady, who has not pleasure in a good novel, must be intolerably stupid."',
 '"Imperfection is beauty, madness is genius and it\'s better to be absolutely ridiculous than absolutely boring."',
 '"Try not to become a man of success. Rather become a man of value."',
 '"It is better to be hated for what you are than to be loved for what you are not."',
 '"I have not failed. I\'ve just found 10,000 ways that won\'t work."',
 '"A woman is like a tea bag; you never know how strong it is until it\'s in hot water."',
 '"A day without sunshine is like, you know, night."']
```

在上述代码中,依据文档中的描述设置参数 url、timeout、images,然后发送 HTTP 请求到服务接口地址。从运行结果看出,页面渲染成功,我们爬取到了页面中的 10 条名人名言。

11.1.2 execute 端点

在爬取某些页面时,我们想在页面中执行一些用户自定义的 JavaScript 代码,例如,用 JavaScript 模拟点击页面中的按钮,或调用页面中的 JavaScript 函数与服务器交互,利用 Splash 的 execute 端点提供的服务可以实现这样的功能。请看表 11-3 中的文档。

表 11-3

服务端点	execute
请求地址	http://localhost:8050/execute
请求方式	POST
返回类型	自定义

execute 端点支持的参数如表 11-4 所示。

表 11-4

参数	必选/可选	类型	描述
lua_source	必选	string	用户自定义的 lua 脚本
timeout	可选	float	渲染页面超时时间
proxy	可选	string	代理服务器地址

我们可以将 execute 端点的服务看作一个可用 lua 语言编程的浏览器，功能类似于 PhantomJS。使用时需传递一个用户自定义的 lua 脚本给 Splash，该 lua 脚本中包含用户想要模拟的浏览器行为，例如：

- 打开某 url 地址的页面
- 等待页面加载及渲染
- 执行 JavaScript 代码
- 获取 HTTP 响应头部
- 获取 Cookie

下面是使用 requests 库调用 execute 端点服务的示例代码。

```
>>> import requests
>>> import json

>>> lua_script = '''
... function main(splash)
...     splash:go("http://example.com")          --打开页面
...     splash:wait(0.5)                          --等待加载
...     local title = splash:evaljs("document.title")   --执行 js 代码获取结果
...     return {title=title}                      --返回 json 形式的结果
... end
... '''
>>> splash_url = 'http://localhost:8050/execute'
>>> headers = {'content-type': 'application/json'}
>>> data = json.dumps({'lua_source': lua_script})
>>> response = requests.post(splash_url, headers=headers, data=data)
>>> response.content
b'{"title": "Example Domain"}'
>>> response.json()
{'title': 'Example Domain'}
```

用户自定义的 lua 脚本中必须包含一个 main 函数作为程序入口，main 函数被调用时会传入一个 splash 对象（lua 中的对象），用户可以调用该对象上的方法操纵 Splash。例如，在上面的例子中，先调用 go 方法打开某页面，再调用 wait 方法等待页面渲染，然后调用 evaljs 方法执行一个 JavaScript 表达式，并将结果转化为相应的 lua 对象，最终 Splash 根据 main 函数的返回值构造 HTTP 响应返回给用户，main 函数的返回值可以是字符串，也可以是 lua 中的表（类似 Python 字典），表会被编码成 json 串。

接下来，看一下 splash 对象常用的属性和方法。

- splash.args 属性

    用户传入参数的表，通过该属性可以访问用户传入的参数，如 splash.args.url、splash.args.wait。

- splash.js_enabled 属性

    用于开启/禁止 JavaScript 渲染，默认为 true。

- splash.images_enabled 属性

    用于开启/禁止图片加载，默认为 true。

- splash:go 方法

    splash:go{url, baseurl=nil, headers=nil, http_method="GET", body=nil, formdata=nil}
    类似于在浏览器中打开某 url 地址的页面，页面所需资源会被加载，并进行 JavaScript 渲染，可以通过参数指定 HTTP 请求头部、请求方法、表单数据等。

- splash:wait 方法

    splash:wait{time, cancel_on_redirect=false, cancel_on_error=true}
    等待页面渲染，time 参数为等待的秒数。

- splash:evaljs 方法

    splash:evaljs(snippet)
    在当前页面下，执行一段 JavaScript 代码，并返回最后一句表达式的值。

- splash:runjs 方法

    splash:runjs(snippet)
    在当前页面下，执行一段 JavaScript 代码，与 evaljs 方法相比，该函数只执行 JavaScript 代码，不返回值。

- splash:url 方法

    splash:url()
    获取当前页面的 url。

- splash:html 方法

    splash:html()
    获取当前页面的 HTML 文本。

- splash:get_cookies 方法

    splash:get_cookies()

    获取全部 Cookie 信息.

11.2 在 Scrapy 中使用 Splash

掌握了 Splash 渲染引擎的基本使用后，我们继续学习如何在 Scrapy 中调用 Splash 服务，Python 库的 scrapy-splash 是非常好的选择。

使用 pip 安装 scrapy-splash：

```
$ pip install scrapy-splash
```

在项目环境中讲解 scrapy-splash 的使用，创建一个 Scrapy 项目，取名为 splash_examples：

```
$ scrapy startproject splash_examples
```

首先在项目配置文件 settings.py 中对 scrapy-splash 进行配置，添加内容如下：

```
# Splash 服务器地址
SPLASH_URL = 'http://localhost:8050'

# 开启 Splash 的两个下载中间件并调整 HttpCompressionMiddleware 的次序
DOWNLOADER_MIDDLEWARES = {
    'scrapy_splash.SplashCookiesMiddleware': 723,
    'scrapy_splash.SplashMiddleware': 725,
    'scrapy.downloadermiddlewares.httpcompression.HttpCompressionMiddleware': 810,
}

# 设置去重过滤器
DUPEFILTER_CLASS = 'scrapy_splash.SplashAwareDupeFilter'

# 用来支持 cache_args（可选）
SPIDER_MIDDLEWARES = {
    'scrapy_splash.SplashDeduplicateArgsMiddleware': 100,
}
```

编写 Spider 代码过程中，使用 scrapy_splash 调用 Splash 服务非常简单，scrapy_splash 中定义了一个 SplashRequest 类，用户只需使用 scrapy_splash.SplashRequest（替代

scrapy.Request）提交请求即可。下面是 SplashRequest 构造器方法中的一些常用参数。

- url
  与 scrapy.Request 中的 url 相同，也就是待爬取页面的 url（注意，不是 Splash 服务器地址）。
- headers
  与 scrapy.Request 中的 headers 相同。
- cookies
  与 scrapy.Request 中的 cookies 相同。
- args
  传递给 Splash 的参数（除 url 以外），如 wait、timeout、images、js_source 等。
- cache_args
  如果 args 中的某些参数每次调用都重复传递并且数据量较大（例如一段 JavaScript 代码），此时可以把该参数名填入 cache_args 列表中，让 Splash 服务器缓存该参数，如 SplashRequest(url, args = {'js_source': js, 'wait': 0.5}, cache_args = ['js_source'])。
- endpoint
  Splash 服务端点，默认为'render.html'，即 JavaScript 页面渲染服务，该参数可以设置为'render.json'、'render.har'、'render.png'、'render.jpeg'、'execute'等，有些服务端点的功能我们没有讲解，详细内容可以查阅文档。
- splash_url
  Splash 服务器地址，默认为 None，即使用配置文件中 SPLASH_URL 的地址。

现在，大家已经对如何在 Scrapy 中使用 Splash 渲染引擎爬取动态页面有了一定了解，接下来我们在已经配置了 Splash 使用环境的 splash_examples 项目中完成两个实战项目。

11.3　项目实战：爬取 toscrape 中的名人名言

11.3.1　项目需求

爬取网站 http://quotes.toscrape.com/js 中的名人名言信息。

11.3.2　页面分析

该网站的页面已在本章开头部分分析过，大家可以回头看相关内容。

11.3.3 编码实现

首先，在 splash_examples 项目目录下使用 scrapy genspider 命令创建 Spider：

```
scrapy genspider quotes quotes.toscrape.com
```

在这个案例中，我们只需使用 Splash 的 render.html 端点渲染页面，再进行爬取即可实现 QuotesSpider，代码如下：

```python
# -*- coding: utf-8 -*-
import scrapy
from scrapy_splash import SplashRequest

class QuotesSpider(scrapy.Spider):
    name = "quotes"
    allowed_domains = ["quotes.toscrape.com"]
    start_urls = ['http://quotes.toscrape.com/js/']

    def start_requests(self):
        for url in self.start_urls:
            yield SplashRequest(url, args={'images': 0, 'timeout': 3})

    def parse(self, response):
        for sel in response.css('div.quote'):
            quote = sel.css('span.text::text').extract_first()
            author = sel.css('small.author::text').extract_first()
            yield {'quote': quote, 'author': author}
        href = response.css('li.next > a::attr(href)').extract_first()
        if href:
            url = response.urljoin(href)
            yield SplashRequest(url, args={'images': 0, 'timeout': 3})
```

上述代码中，使用 SplashRequest 提交请求，在 SplashRequest 的构造器中无须传递 endpoint 参数，因为该参数默认值便是'render.html'。使用 args 参数禁止 Splash 加载图片，并设置渲染超时时间。

运行爬虫，观察结果：

```
$ scrapy crawl quotes -o quotes.csv
...
```

```
$ cat -n quotes.csv
     1  quote,author
     2  "The world as we have created it is a process of our thinking. It cannot be changed without changing our thinking.",Albert Einstein
     3  ""It is our choices, Harry, that show what we truly are, far more than our abilities."",J.K. Rowling
     4  "There are only two ways to live your life. One is as though nothing is a miracle. The other is as though everything is a miracle.",Albert Einstein
     5  ""The person, be it gentleman or lady, who has not pleasure in a good novel, must be intolerably stupid."",Jane Austen
     6  ""Imperfection is beauty, madness is genius and it's better to be absolutely ridiculous than absolutely boring."",Marilyn Monroe
     7  "Try not to become a man of success. Rather become a man of value.",Albert Einstein
     8  "It is better to be hated for what you are than to be loved for what you are not.",André Gide
     9  ""I have not failed. I've just found 10,000 ways that won't work."",Thomas A. Edison
    10  "A woman is like a tea bag; you never know how strong it is until it's in hot water.",Eleanor Roosevelt
    ...
    91  ""I believe in Christianity as I believe that the sun has risen: not only because I see it, but because by it I see everything else."",C.S. Lewis
    92  ""The truth."" Dumbledore sighed. ""It is a beautiful and terrible thing, and should therefore be treated with great caution."",J.K. Rowling
    93  ""I'm the one that's got to die when it's time for me to die, so let me live my life the way I want to."",Jimi Hendrix
    94  "To die will be an awfully big adventure.",J.M. Barrie
    95  "It takes courage to grow up and become who you really are.",E.E. Cummings
    96  "But better to get hurt by the truth than comforted with a lie.",Khaled Hosseini
    97  "You never really understand a person until you consider things from his point of view... Until you climb inside of his skin and walk around in it.",Harper Lee
    98  ""You have to write the book that wants to be written. And if the book will be too difficult for grown-ups, then you write it for children."",Madeleine L'Engle
    99  "Never tell the truth to people who are not worthy of it.",Mark Twain
   100  ""A person's a person, no matter how small."",Dr. Seuss
   101  ""... a mind needs books as a sword needs a whetstone, if it is to keep its edge."",George R.R. Martin
```

运行结果显示，我们成功爬取了 10 个页面中的 100 条名人名言。

11.4 项目实战：爬取京东商城中的书籍信息

11.4.1 项目需求

爬取京东商城中所有 Python 书籍的名字和价格信息。

11.4.2 页面分析

图 11-4 所示为在京东网站（http://www.jd.com）的书籍分类下搜索 Python 关键字得到的页面。

图 11-4

结果有很多页，在每一个书籍列表页面中可以数出有 60 本书，但我们在 scrapy shell 中爬取该页面时遇到了问题，仅在页面中找到了 30 本书，少了 30 本，代码如下：

```
$ scrapy shell
...
>>> url = 'https://search.jd.com/Search?keyword=python&enc=utf-8&book=y&wq=python'
>>> fetch(url)
...
```

```
>>> len(response.css('ul.gl-warp > li'))
30
```

原来页面中的 60 本书不是同时加载的,开始只有 30 本书,当我们使用鼠标滚轮滚动到页面下方某位置时,后 30 本书才由 JavaScript 脚本加载,通过实验可以验证这个说法,实验流程如下:

(1)页面刚加载时,在 Chrome 开发者工具的 console 中用 jQuery 代码查看当前有多少本书,此时为 30。

(2)之后滚动鼠标滚轮到某一位置时,可以看到 JavaScript 发送 HTTP 请求和服务器交互(XHR)。

(3)然后用 jQuery 代码查看当前有多少本书,已经变成了 60,如图 11-5 所示。

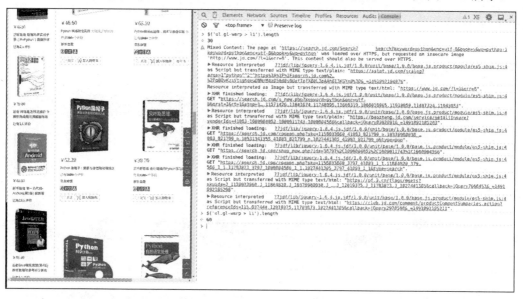

图 11-5

既然如此,爬取这个页面时,可以先执行一段 JavaScript 代码,将滚动条拖到页面下方某位置,触发加载后 30 本书的事件,在开发者工具的 console 中进行实验,用 document.getElementsByXXX 方法随意选中页面下方的某元素,比如"下一页"按钮所在的<div>元素,然后在该元素对象上调用 scrollIntoView(true)完成拖曳动作,此时查看书籍数量,变成了 60,这个解决方案是可行的。图 11-6 所示为实验过程。

爬取一个页面的问题解决了,再来研究如何从页面中找到下一页的 url 地址。

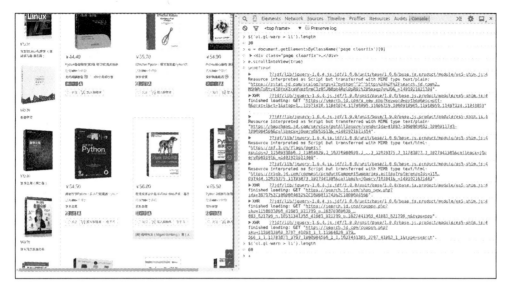

图 11-6

如图 11-7 所示,下一页链接的 href 属性并不是一个 url,而在其 onclick 属性中包含了一条 JavaScript 代码,单击"下一页"按钮时会调用函数 SEARCH.page(n, true)。虽然可以用 Splash 执行函数来跳转到下一页,但还是很麻烦,经观察发现,每个页面 url 的差异仅在于 page 参数不同,第一页 page=1,第 2 页 page=3,第 3 页 page=5……以 2 递增,并且页面中还包含商品总数信息。因此,我们可以推算出所有页面的 url。

图 11-7

11.4.3 编码实现

首先，在 splash_examples 项目目录下使用 scrapy genspider 命令创建 Spider 类：

```
scrapy genspider jd_book search.jd.com
```

经上述分析，在爬取每一个书籍列表页面时都需要执行一段 JavaScript 代码，以让全部书籍加载，因此选用 execute 端点完成该任务，实现 JDBookSpider 代码如下：

```python
# -*- coding: utf-8 -*-
import scrapy
from scrapy import Request
from scrapy_splash import SplashRequest

lua_script = '''
function main(splash)
    splash:go(splash.args.url)
    splash:wait(2)
    splash:runjs("document.getElementsByClassName('page')[0].scrollIntoView(true)")
    splash:wait(2)
    return splash:html()
end
'''

class JDBookSpider(scrapy.Spider):
    name = "jd_book"
    allowed_domains = ["search.jd.com"]
    base_url = 'https://search.jd.com/Search?keyword=python&enc=utf-8&book=y&wq=python'

    def start_requests(self):
        # 请求第一页，无须 js 渲染
        yield Request(self.base_url, callback=self.parse_urls, dont_filter=True)

    def parse_urls(self, response):
        # 获取商品总数，计算出总页数
        total = int(response.css('span#J_resCount::text').extract_first())
        pageNum = total // 60 + (1 if total % 60 else 0)

        # 构造每页的 url，向 Splash 的 execute 端点发送请求
        for i in range(pageNum):
            url = '%s&page=%s' % (self.base_url, 2 * i + 1)
            yield SplashRequest(url,
```

```
                    endpoint='execute',
                    args={'lua_source': lua_script},
                    cache_args=['lua_source'])

    def parse(self, response):
        # 获取一个页面中每本书的名字和价格
        for sel in response.css('ul.gl-warp.clearfix > li.gl-item'):
            yield {
                'name': sel.css('div.p-name').xpath('string(.//em)').extract_first(),
                'price': sel.css('div.p-price i::text').extract_first(),
            }
```

解释上述代码如下：

- start_requests 方法
  start_requests 提交对第一个页面的请求，这个页面不需要渲染，因为我们只想从中获取页面总数，使用 scrapy.Request 提交请求，并指定 parse_urls 作为解析函数。
- parse_urls 方法
  从第一个页面中提取商品总数，用其计算页面总数，之后按照前面分析出的页面 url 规律构造每一个页面的 url。这些页面都是需要渲染的，使用 SplashRequest 提交请求，除了渲染页面以外，还需要执行一段 JavaScript 代码（为了加载后 30 本书），因此使用 Splash 的 execute 端点将 endpoint 参数置为'execute'。通过 args 参数的 lua_source 字段传递我们要执行的 lua 脚本，由于爬取每个页面时都要执行该脚本，因此可以使用 cache_args 参数将该脚本缓存到 Splash 服务器。
- parse 方法
  一个页面中提取 60 本书的名字和价格信息，相关内容大家早已熟悉，不再赘述。
- lua_script 字符串
  自定义的 lua 脚本，其中的逻辑很简单：
  打开页面→等待渲染→执行 js 触发数据加载（后 30 本书）→等待渲染→返回 html。

另外，京东服务器程序会对请求头部中的 User-Agent 字段进行检测，因此需要在配置文件 settings.py 中设置 USER_AGENT，伪装成常规浏览器：

```
# 伪装成常规浏览器
USER_AGENT = 'Mozilla/5.0 (X11; Linux x86_64) AppleWebKit/537.36 (KHTML, like Gecko)'
```

编码和配置的工作已经完成了，运行爬虫并观察结果：

```
$ scrapy crawl jd_book -o books.csv
...
$ cat -n books.scv
     1  name,price
     2  教孩子学编程 Python 语言版,59.00
     3  Python 机器学习 预测分析核心算法,46.90
     4  Python 机器学习 预测分析核心算法 9787115433732 [美] Michael,52.44
     5  Python 数据分析实战 9787115432209,39.50
     6  数据科学实战手册 R+Python 9787115426758,39.50
     7  Python 编程 从入门到实践,89.00
     8  Python 可以这样学,69.00
     9  Python 游戏编程快速上手,47.20
    10  Python 性能分析与优化,45.00
   ...
  2932  Python 语言在 Abaqus 中的应用（附 CD-ROM 光盘 1 张）,38.40
  2933  Python 绝技 运用 Python 成为*级黑客 运用 Python 成为 级黑客  书籍,46.50
  2934  趣学 Python 编程,42.70
  2935  零基础学 Python（图文版）,65.50
  2936  Python 程序设计入门到实战,65.50
  2937  计算机科学丛书：Python 语言程序设计,74.60
  2938  面向 ArcGIS 的 Python 脚本编程,41.00
  2939  Python 算法教程  书籍,40.60
  2940  Python 基础教程+利用 Python 进行数据分析 Python 学习套装 共两册,136.20
  2941  包邮 量化投资 以 Python 为工具 Python 语言处理数据 Python 金融 量化投,65.80
```

结果显示，我们成功爬取到了 2940 本书籍的信息。

11.5 本章小结

本章学习了爬取动态页面的相关知识，首先带大家了解了动态页面的实现原理，然后介绍了页面渲染引擎 Splash，并详细讲解了其中 render.html 和 execute 两个服务端点的使用，最后通过两个案例展示了在 Scrapy 爬虫中如何利用 Splash 服务爬取动态页面的内容。

第 12 章

存入数据库

在之前的章节中,曾讨论过将爬取到的数据导出到文件的相关话题,但在某些时候,我们希望将爬取到的数据存储到数据库中,这一章来学习使用 Item Pipeline 实现 Scrapy 爬虫和几种常用数据库的接口。

以第 8 章 toscrape_book 项目作为环境展开本章内容的讲解。在 toscrape_book 项目中,我们爬取了网站 http://books.toscrape.com 中的书籍信息,其中每一本书的信息包括:

- 书名
- 价格
- 评价等级
- 产品编码
- 库存量
- 评价数量

下面我们来学习如何在爬取数据的过程中将书籍信息存储到各种数据库,这些数据库主要有:SQLite、MySQL、MongoDB、Redis。

12.1 SQLite

SQLite 是一个文件型轻量级数据库，它的处理速度很快，在数据量不是很大的情况下，使用 SQLite 足够了。

首先，创建一个供 Scrapy 使用的 SQLite 数据库，取名为 scrapy.db：

```
$ sqlite3 scrapy.db
...
sqlite>
```

接下来，在客户端中创建数据表（Table）：

```
CREATE TABLE books (
    upc             CHAR(16) NOT NULL PRIMARY KEY,
    name            VARCHAR(256) NOT NULL,
    price           VARCHAR(16) NOT NULL,
    review_rating   INT,
    review_num      INT,
    stock           INT
);
```

在 Python 中访问 SQLite 数据库可使用 Python 标准库中的 sqlite3 模块。下面是使用 sqlite3 模块将数据写入 SQLite 数据库的简单示例：

```
import sqlite3

#连接数据库，得到 Connection 对象
conn = sqlite3.connect('example.db')

#创建 Curosr 对象，用来执行 SQL 语句
cur = conn.cursor()

#创建数据表
cur.execute('CREATE TABLE person (name VARCHAR(32), age INT, sex char(1))')

#插入一条数据
cur.execute('INSERT INTO person VALUES (?,?,?)', ('刘硕', 34, 'M'))
```

```python
#保存变更,commit 后数据才被实际写入数据库
conn.commit()

#关闭连接
conn.close()
```

了解了在 Python 中如何操作 SQLite 数据库后,接下来编写一个能将爬取到的数据写入 SQLite 数据库的 Item Pipeline。在 pipelines.py 中实现 SQLitePipeline 的 代码如下:

```python
import sqlite3

class SQLitePipeline(object):

    def open_spider(self, spider):
        db_name = spider.settings.get('SQLITE_DB_NAME', 'scrapy_defaut.db')

        self.db_conn = sqlite3.connect(db_name)
        self.db_cur = self.db_conn.cursor()

    def close_spider(self, spider):
        self.db_conn.commit()
        self.db_conn.close()

    def process_item(self, item, spider):
        self.insert_db(item)

        return item

    def insert_db(self, item):
        values = (
            item['upc'],
            item['name'],
            item['price'],
            item['review_rating'],
            item['review_num'],
            item['stock'],
        )

        sql = 'INSERT INTO books VALUES (?,?,?,?,?,?)'
```

```
            self.db_cur.execute(sql, values)

            # 每插入一条就 commit 一次会影响效率
            # self.db_conn.commit()
```

解释上述代码如下：

- open_spider 方法在开始爬取数据之前被调用，在该方法中通过 spider.settings 对象读取用户在配置文件中指定的数据库，然后建立与数据库的连接，将得到的 Connection 对象和 Cursor 对象分别赋值给 self.db_conn 和 self.db_cur，以便之后使用。
- process_item 方法处理爬取到的每一项数据，在该方法中调用 insert_db 方法，执行插入数据操作的 SQL 语句。但需要注意的是，在 insert_db 中并没有调用连接对象的 commit 方法，也就意味着此时数据并没有实际写入数据库。如果每插入一条数据都调用一次 commit 方法，会严重降低程序执行效率，并且我们对数据插入数据库的实时性并没有什么要求，因此可以在爬取完全部数据后再调用 commit 方法。
- close_spider 方法在爬取完全部数据后被调用，在该方法中，调用连接对象的 commit 方法将之前所有的插入数据操作一次性提交给数据库，然后关闭连接对象。

在配置文件 settings.py 中指定我们所要使用的 SQLite 数据库，并启用 SQLitePipeline：

```
SQLITE_DB_NAME = 'scrapy.db'

ITEM_PIPELINES = {
    'toscrape_book.pipelines.SQLitePipeline': 400,
}
```

运行爬虫，并查看数据库：

```
$ scrapy crawl books
...
$ sqlite3 scrapy.db
SQLite version 3.8.2 2013-12-06 14:53:30
Enter ".help" for instructions
Enter SQL statements terminated with a ";"
sqlite> select count(*) from books;
1000
sqlite> select * from books;
a22124811bfa8350|It's Only the Himalayas|£45.17|2|0|19
feb7cc7701ecf901|Olio|£23.88|1|0|19
```

```
a34ba96d4081e6a4|Rip it Up and Start Again|£35.02|5|0|19
a18a4f574854aced|Libertarianism for Beginners|£51.33|2|0|19
ce6396b0f23f6ecc|Set Me Free|£17.46|5|0|19
fa9610a50a1bf149|Masks and Shadows|£56.40|2|0|16
3c346ab1e76ae1f6|Obsidian (Lux #1)|£14.86|2|0|16
09b6cc87e62c2c58|Danganronpa Volume 1|£51.99|4|0|16
...
```

结果表明，我们成功地将 1000 条数据存储到了 SQLite 数据库。

12.2 MySQL

MySQL 是一个应用极其广泛的关系型数据库，它是开源免费的，可以支持大型数据库，在个人用户和中小企业中成为技术首选。

使用客户端登录 MySQL，创建一个供 Scrapy 使用的数据库，取名为 scrapy_db：

```
$ mysql -hlocalhost -uliushuo -p12345678
...
mysql> CREATE DATABASE scrapy_db CHARACTER SET 'utf8' COLLATE 'utf8_general_ci';
Query OK, 1 row affected (0.00 sec)
mysql> USE scrapy_db;
Database changed
```

接下来，创建存储书籍数据的表：

```
mysql> CREATE TABLE books (
    ->     upc             CHAR(16) NOT NULL PRIMARY KEY,
    ->     name            VARCHAR(256) NOT NULL,
    ->     price           VARCHAR(16) NOT NULL,
    ->     review_rating   INT,
    ->     review_num      INT,
    ->     stock           INT
    -> ) ENGINE=InnoDB DEFAULT CHARSET=utf8;
Query OK, 0 rows affected (0.08 sec)
```

在 Python 2 中访问 MySQL 数据库可以使用第三方库 MySQL-Python（即 MySQLdb），但是 MySQLdb 不支持 Python 3。在 Python 3 中，可以使用另一个第三方库 mysqlclient 作为替代，它是基于 MySQL-Python 开发的，提供了几乎完全相同的接口。因此，在两个 Python 版本下，可以使用相同的代码访问 MySQL。

Python 2 使用 pip 安装 MySQL-python：

```
sudo pip install MySQL-python
```

Python 3 使用 pip 安装 mysqlclient：

```
sudo pip install mysqlclient
```

下面是使用 MySQLdb 将数据写入 MySQL 数据库的简单示例，与 sqlite3 的使用几乎完全相同：

```python
import MySQLdb

#连接数据库，得到 Connection 对象
conn = MySQLdb.connect(host='localhost', db='scrapy_db',
            user='liushuo', passwd='12345678', charset='utf8')

#创建 Curosr 对象，用来执行 SQL 语句
cur = conn.cursor()

#创建数据表
cur.execute('CREATE TABLE person (name VARCHAR(32), age INT, sex char(1)) \
            ENGINE=InnoDB DEFAULT CHARSET=utf8')

#插入一条数据
cur.execute('INSERT INTO person VALUES (%s,%s,%s)', ('刘硕', 34, 'M'))

#保存变更，commit 后数据才被实际写入数据库
conn.commit()

#关闭连接
conn.close()
```

仿照 SQLitePipeline 实现 MySQLPipeline，代码如下：

```python
import MySQLdb

class MySQLPipeline:
    def open_spider(self, spider):
        db = spider.settings.get('MYSQL_DB_NAME', 'scrapy_default')
        host = spider.settings.get('MYSQL_HOST', 'localhost')
```

```python
            port = spider.settings.get('MYSQL_PORT', 3306)
            user = spider.settings.get('MYSQL_USER', 'root')
            passwd = spider.settings.get('MYSQL_PASSWORD', 'root')

            self.db_conn = MySQLdb.connect(host=host, port=port, db=db,
                                    user=user, passwd=passwd, charset='utf8')
            self.db_cur = self.db_conn.cursor()

        def close_spider(self, spider):
            self.db_conn.commit()
            self.db_conn.close()

        def process_item(self, item, spider):
            self.insert_db(item)

            return item

        def insert_db(self, item):
            values = (
                item['upc'],
                item['name'],
                item['price'],
                item['review_rating'],
                item['review_num'],
                item['stock'],
            )

            sql = 'INSERT INTO books VALUES (%s,%s,%s,%s,%s,%s)'
            self.db_cur.execute(sql, values)
```

上述代码结构与 SQLitePipeline 完全相同，不再赘述。

在配置文件 settings.py 中指定我们所要使用的 MySQL 数据库，并启用 MySQLPipeline：

```
MYSQL_DB_NAME = 'scrapy_db'
MYSQL_HOST = 'localhost'
MYSQL_USER = 'liushuo'
MYSQL_PASSWORD = '12345678'

ITEM_PIPELINES = {
```

```
        'toscrape_book.pipelines.MySQLPipeline': 401,
}
```

运行爬虫，并查看数据库：

```
$ scrapy crawl books
...
$ mysql -hlocalhost -uliushuo -p12345678 scrapy_db
...
mysql> select count(*) from books;
+----------+
| count(*) |
+----------+
|     1000 |
+----------+
1 row in set (0.00 sec)
mysql> select name from books;
+------------------------------------------------------------------+
| name                                                             |
+------------------------------------------------------------------+
| Deep Under (Walker Security #1)                                  |
| Black Dust                                                       |
| The Songs of the Gods                                            |
| The Passion of Dolssa                                            |
| The Rosie Project (Don Tillman #1)                               |
| Give It Back                                                     |
| A Brush of Wings (Angels Walking #3)                             |
| A Court of Thorns and Roses (A Court of Thorns and Roses #1)     |
| You (You #1)                                                     |
| Travels with Charley: In Search of America                       |
| Old School (Diary of a Wimpy Kid #10)                            |
| Starving Hearts (Triangular Trade Trilogy, #1)                   |
| The Origin of Species                                            |
| Is Everyone Hanging Out Without Me? (And Other Concerns)         |
| A Game of Thrones (A Song of Ice and Fire #1)                    |
| Living Forward: A Proven Plan to Stop Drifting and Get the Life  |
```

```
...省略中间输出...

| The Perks of Being a Wallflower                          |
| The Vacationers                                          |
| My Perfect Mistake (Over the Top #1)                     |
| I Had a Nice Time And Other Lies...: How to find love    |
| In Cold Blood                                            |
| Still Life with Bread Crumbs                             |
| The Expatriates                                          |
| Soldier (Talon #3)                                       |
| Olio                                                     |
| The Dream Thieves (The Raven Cycle #2)                   |
| A Murder Over a Girl: Justice, Gender, Junior High       |
| The Demonists (Demonist #1)                              |
| The Husband's Secret                                     |
| Amatus                                                   |
| Art Ops Vol. 1                                           |
| The Haters                                               |
+----------------------------------------------------------+
```

结果表明，我们成功地将 1000 条数据存储到了 MySQL 数据库。

上述代码中，同样是先执行完全部的插入语句（INSERT INTO），最后一次性调用 commit 方法提交给数据库。或许在某些情况下，我们的确需要每执行一条插入语句，就立即调用 commit 方法更新数据库，如爬取过程很长，中途可能被迫中断，这样程序就不能执行到最后的 commit。如果在上述代码的 insert_db 方法中直接添加 self.db_conn.commit()，又会使程序执行慢得让人无法忍受。为解决以上难题，下面讲解另一种实现方法。

Scrapy 框架自身是使用另一个 Python 框架 Twisted 编写的程序，Twisted 是一个事件驱动型的异步网络框架，鼓励用户编写异步代码，Twisted 中提供了以异步方式多线程访问数据库的模块 adbapi，使用该模块可以显著提高程序访问数据库的效率。下面是使用 adbapi 中的连接池访问 MySQL 数据库的简单示例：

```
from twisted.internet import    reactor, defer
from twisted.enterprise import adbapi
import threading

dbpool = adbapi.ConnectionPool('MySQLdb', host='localhost', database='scrapy_db',
                    user='liushuo', password='liushuo', charset='utf8')
```

```python
def insert_db(tx, item):
    print('In Thread:', threading.get_ident())
    sql = 'INSERT INTO person VALUES (%s, %s, %s)'
    tx.execute(sql, item)

for i in range(1000):
    item = ('person%s' % i, 25, 'M')
    dbpool.runInteraction(insert_db, item)

reactor.run()
```

上述代码解释如下:

- adbapi.ConnectionPool 方法可以创建一个数据库连接池对象，其中包含多个连接对象，每个连接对象在独立的线程中工作。adbapi 只是提供了异步访问数据库的编程框架，在其内部依然使用 MySQLdb、sqlite3 这样的库访问数据库。ConnectionPool 方法的第一个参数就是用来指定使用哪个库访问数据库，其他参数在创建连接对象时使用。

- dbpool.runInteraction(insert_db, item) 以异步方式调用 insert_db 函数，dbpool 会选择连接池中的一个连接对象在独立线程中调用 insert_db，其中参数 item 会被传给 insert_db 的第二个参数，传给 insert_db 的第一个参数是一个 Transaction 对象，其接口与 Cursor 对象类似，可以调用 execute 方法执行 SQL 语句，insert_db 执行完后，连接对象会自动调用 commit 方法。

了解了 adbapi 的使用后，给出第二个版本的 MySQLPipeline，代码如下:

```python
from twisted.enterprise import adbapi

class MySQLAsyncPipeline:
    def open_spider(self, spider):
        db = spider.settings.get('MYSQL_DB_NAME', 'scrapy_default')
        host = spider.settings.get('MYSQL_HOST', 'localhost')
        port = spider.settings.get('MYSQL_PORT', 3306)
        user = spider.settings.get('MYSQL_USER', 'root')
        passwd = spider.settings.get('MYSQL_PASSWORD', 'root')

        self.dbpool = adbapi.ConnectionPool('MySQLdb', host=host, db=db,
                                             user=user, passwd=passwd, charset='utf8')
```

```
def close_spider(self, spider):
    self.dbpool.close()

def process_item(self, item, spider):
    self.dbpool.runInteraction(self.insert_db, item)

    return item

def insert_db(self, tx, item):
    values = (
        item['upc'],
        item['name'],
        item['price'],
        item['review_rating'],
        item['review_num'],
        item['stock'],
    )

    sql = 'INSERT INTO books VALUES (%s,%s,%s,%s,%s,%s)'
    tx.execute(sql, values)
```

通过前面的讲述，相信大家可以轻松理解上述代码，不再过多解释，该版本比之前的版本在执行效率上有显著提高。

12.3 MongoDB

MongoDB 是一个面向文档的非关系型数据库（NoSQL），它功能强大、灵活、易于拓展，近年来在多个领域得到广泛应用。

在 Python 中可以使用第三方库 pymongo 访问 MongoDB 数据库，使用 pip 安装 pymongo：

```
$ sudo pip install pymongo
```

下面是使用 pymongo 将数据写入 MongoDB 数据库的简单示例：

```
from pymongo import MongoClient
```

```python
# 连接MongoDB,得到一个客户端对象
client = MongoClient('mongodb://localhost:27017')

# 获取名为 scrapy_db 的数据库的对象
db = client.scrapy_db

# 获取名为 person 的集合的对象
collection = db.person

doc = {
    'name': '刘硕',
    'age': 34,
    'sex': 'M',
}

# 将文档插入集合
collection.insert_one(doc)

# 关闭客户端
client.close()
```

仿照 SQLitePipeline 实现 MongoDBPipeline,代码如下:

```python
from pymongo import MongoClient
from scrapy import Item

class MongoDBPipeline:
    def open_spider(self, spider):
        db_uri = spider.settings.get('MONGODB_URI', 'mongodb://localhost:27017')
        db_name = spider.settings.get('MONGODB_DB_NAME', 'scrapy_default')

        self.db_client = MongoClient('mongodb://localhost:27017')
        self.db = self.db_client[db_name]

    def close_spider(self, spider):
        self.db_client.close()

    def process_item(self, item, spider):
        self.insert_db(item)
```

```
            return item

        def insert_db(self, item):
            if isinstance(item, Item):
                item = dict(item)

            self.db.books.insert_one(item)
```

解释上述代码如下：

- open_spider 方法在开始爬取数据之前被调用，在该方法中通过 spider.settings 对象读取用户在配置文件中指定的数据库，然后建立与数据库的连接，将得到的 MongoClient 对象和 Database 对象分别赋值给 self.db_client 和 self.db，以便之后使用。
- process_item 方法处理爬取到的每一项数据，在该方法中调用 insert_db 方法，执行数据库的插入操作。在 insert_db 方法中，先将一项数据转换成字典，然后调用 insert_one 方法将其插入集合 books。
- close_spider 方法在爬取完全部数据后被调用，在该方法中关闭与数据库的连接。

在配置文件 settings.py 中指定我们所要使用的 MongoDB 数据库，并启用 MongoDBPipeline：

```
MONGODB_URI = 'mongodb://localhost:27017'
MONGODB_DB_NAME = 'scrapy_db'

ITEM_PIPELINES = {
    'toscrape_book.pipelines.MongoDBPipeline': 403,
}
```

运行爬虫，并查看数据库：

```
$ scrapy crawl books
...
$ mongo scrapy_db
MongoDB shell version: 2.4.9
connecting to: scrapy_db
> db.books.count()
1000
> db.books.find()
```

{ "_id" : ObjectId("58fb48859dcd1928b736ee4f"), "review_rating" : 3, "review_num" : "0", "stock" : "22", "upc" : "a897fe39b1053632", "price" : "£51.77", "name" : "A Light in the Attic" }

{ "_id" : ObjectId("58fb48859dcd1928b736ee50"), "review_rating" : 1, "review_num" : "0", "stock" : "19", "upc" : "feb7cc7701ecf901", "price" : "£23.88", "name" : "Olio" }

{ "_id" : ObjectId("58fb48859dcd1928b736ee51"), "review_rating" : 2, "review_num" : "0", "stock" : "19", "upc" : "a18a4f574854aced", "price" : "£51.33", "name" : "Libertarianism for Beginners" }

{ "_id" : ObjectId("58fb48859dcd1928b736ee52"), "review_rating" : 1, "review_num" : "0", "stock" : "19", "upc" : "e30f54cea9b38190", "price" : "£37.59", "name" : "Mesaerion: The Best Science Fiction Stories 1800-1849" }

{ "_id" : ObjectId("58fb48859dcd1928b736ee53"), "review_rating" : 5, "review_num" : "0", "stock" : "19", "upc" : "a34ba96d4081e6a4", "price" : "£35.02", "name" : "Rip it Up and Start Again" }

{ "_id" : ObjectId("58fb48859dcd1928b736ee54"), "review_rating" : 2, "review_num" : "0", "stock" : "19", "upc" : "a22124811bfa8350", "price" : "£45.17", "name" : "It's Only the Himalayas" }

{ "_id" : ObjectId("58fb48859dcd1928b736ee55"), "review_rating" : 5, "review_num" : "0", "stock" : "19", "upc" : "3b1c02bac2a429e6", "price" : "£52.29", "name" : "Scott Pilgrim's Precious Little Life (Scott Pilgrim #1)" }

{ "_id" : ObjectId("58fb48859dcd1928b736ee56"), "review_rating" : 3, "review_num" : "0", "stock" : "19", "upc" : "deda3e61b9514b83", "price" : "£57.25", "name" : "Our Band Could Be Your Life: Scenes from the American Indie Underground, 1981-1991" }

{ "_id" : ObjectId("58fb48869dcd1928b736ee57"), "review_rating" : 5, "review_num" : "0", "stock" : "19", "upc" : "ce6396b0f23f6ecc", "price" : "£17.46", "name" : "Set Me Free" }

{ "_id" : ObjectId("58fb48869dcd1928b736ee58"), "review_rating" : 4, "review_num" : "0", "stock" : "19", "upc" : "30a7f60cd76ca58c", "price" : "£20.66", "name" : "Shakespeare's Sonnets" }

{ "_id" : ObjectId("58fb48869dcd1928b736ee59"), "review_rating" : 2, "review_num" : "0", "stock" : "19", "upc" : "0312262ecafa5a40", "price" : "£13.99", "name" : "Starving Hearts (Triangular Trade Trilogy, #1)" }

{ "_id" : ObjectId("58fb48869dcd1928b736ee5a"), "review_rating" : 1, "review_num" : "0", "stock" : "19", "upc" : "1dfe412b8ac00530", "price" : "£52.15", "name" : "The Black Maria" }

{ "_id" : ObjectId("58fb48869dcd1928b736ee5b"), "review_rating" : 4, "review_num" : "0", "stock" : "19", "upc" : "e10e1e165dc8be4a", "price" : "£22.60", "name" : "The Boys in the Boat: Nine Americans and Their Epic Quest for Gold at the 1936 Berlin Olympics" }

{ "_id" : ObjectId("58fb48869dcd1928b736ee5c"), "review_rating" : 1, "review_num" : "0", "stock" : "19", "upc" : "f77dbf2323deb740", "price" : "£22.65", "name" : "The Requiem Red" }

{ "_id" : ObjectId("58fb48869dcd1928b736ee5d"), "review_rating" : 4, "review_num" : "0", "stock" : "19", "upc" : "2597b5a345f45e1b", "price" : "£33.34", "name" : "The Dirty Little Secrets of Getting Your Dream Job" }

{ "_id" : ObjectId("58fb48869dcd1928b736ee5e"), "review_rating" : 3, "review_num" : "0", "stock" : "19", "upc" : "e72a5dfc7e9267b2", "price" : "£17.93", "name" : "The Coming Woman: A Novel Based on

the Life of the Infamous Feminist, Victoria Woodhull" }
    { "_id" : ObjectId("58fb48869dcd1928b736ee5f"), "review_rating" : 5, "review_num" : "0", "stock" : "20", "upc" : "4165285e1663650f", "price" : "£54.23", "name" : "Sapiens: A Brief History of Humankind" }
    { "_id" : ObjectId("58fb48869dcd1928b736ee60"), "review_rating" : 4, "review_num" : "0", "stock" : "20", "upc" : "e00eb4fd7b871a48", "price" : "£47.82", "name" : "Sharp Objects" }
    { "_id" : ObjectId("58fb48869dcd1928b736ee61"), "review_rating" : 1, "review_num" : "0", "stock" : "20", "upc" : "90fa61229261140a", "price" : "£53.74", "name" : "Tipping the Velvet" }
    { "_id" : ObjectId("58fb48869dcd1928b736ee62"), "review_rating" : 1, "review_num" : "0", "stock" : "20", "upc" : "6957f44c3847a760", "price" : "£50.10", "name" : "Soumission" }
    Type "it" for more
>
```

结果表明，我们成功地将 1000 条数据存储到了 MongoDB 数据库。

## 12.4 Redis

Redis 是一个使用 ANSI C 编写的高性能 Key-Value 数据库，使用内存作为主存储，内存中的数据也可以被持久化到硬盘。

在 Python 中可以使用第三方库 redis-py 访问 Redis 数据库，使用 pip 安装 redis-py：

```
$ sudo pip install redis
```

下面是使用 redis-py 将数据写入 Redis 数据库的简单示例：

```python
import redis

# 连接数据库
r = redis.StrictRedis(host='localhost', port=6379, db=0)

# 创建 3 条数据
person1 = {
    'name': '刘硕',
    'age': 34,
    'sex': 'M',
}

person2= {
    'name': '李雷',
    'age': 32,
```
```

```
        'sex': 'M',
}

person3= {
    'name': '韩梅梅',
    'age': 31,
    'sex': 'F',
}

# 将 3 条数据以 Hash 类型（哈希）保存到 Redis 中
r.hmset('person:1', person1)
r.hmset('person:2', person2)
r.hmset('person:3', person3)

# 关闭连接
r.connection_pool.disconnect()
```

Redis 是 Key-Value 数据库，一项数据在数据库中就是一个键值对，存储多项同类别的数据时（如 Book），通常以 item:id 这样的形式作为每项数据的键，其中的"："并没有什么特殊，也可以换成"-"或"/"等，只是大家习惯这样使用。

仿照 SQLitePipeline 实现 RedisPipeline，代码如下：

```
import redis
from scrapy import Item

class RedisPipeline:
    def open_spider(self, spider):
        db_host = spider.settings.get('REDIS_HOST', 'localhost')
        db_port = spider.settings.get('REDIS_PORT', 6379)
        db_index = spider.settings.get('REDIS_DB_INDEX', 0)

        self.db_conn = redis.StrictRedis(host=db_host, port=db_port, db=db_index)
        self.item_i = 0

    def close_spider(self, spider):
        self.db_conn.connection_pool.disconnect()
```

```
        def process_item(self, item, spider):
            self.insert_db(item)

            return item

        def insert_db(self, item):
            if isinstance(item, Item):
                item = dict(item)

            self.item_i += 1
            self.db_conn.hmset('book:%s' % self.item_i, item)
```

解释上述代码如下：

- open_spider 方法在开始爬取数据之前被调用，在该方法中通过 spider.settings 对象读取用户在配置文件中指定的数据库，然后建立与数据库的连接，将得到的连接对象赋值给 self.db_conn，以便之后使用，并初始化一个 self.item_i 作为每项数据的 id。在插入一项数据时，使用 self.item_i 自加的结果构造数据在数据库中的键。
- process_item 方法处理爬取到的每一项数据，在该方法中调用 insert_db 方法执行数据库的插入操作，在 insert_db 方法中先将一项数据转换成字典，然后调用 hmset 方法将数据以 Hash 类型存入 Redis 数据库。
- close_spider 方法在爬取完全部数据后被调用，在该方法中关闭与数据库的连接。

在配置文件 settings.py 中指定我们所要使用的 Redis 数据库，并启用 RedisPipeline：

```
REDIS_HOST = 'localhost'
REDIS_PORT = 6379
REDIS_DB_INDEX = 0

ITEM_PIPELINES = {
  'toscrape_book.pipelines.RedisPipeline': 404,
}
```

运行爬虫，并查看数据库：

```
$ scrapy crawl books
$ redis-cli
127.0.0.1:6379> KEYS book:*
  1) "book:470"
```

2) "book:300"
  3) "book:801"
  4) "book:476"
  5) "book:914"

...省略中间输出...

 995) "book:703"
 996) "book:407"
 997) "book:995"
 998) "book:569"
 999) "book:298"
1000) "book:110"
127.0.0.1:6379> HGETALL book:1
 1) "price"
 2) "\xc2\xa351.33"
 3) "review_rating"
 4) "2"
 5) "review_num"
 6) "0"
 7) "name"
 8) "Libertarianism for Beginners"
 9) "stock"
10) "19"
11) "upc"
12) "a18a4f574854aced"
127.0.0.1:6379> HGETALL book:2
 1) "price"
 2) "\xc2\xa317.46"
 3) "review_rating"
 4) "5"
 5) "review_num"
 6) "0"
 7) "name"
 8) "Set Me Free"
 9) "stock"
10) "19"

11) "upc"
12) "ce6396b0f23f6ecc"

结果表明，我们成功地将 1000 条数据存储到了 Redis 数据库。

12.5 本章小结

本章学习了如何将爬取到的数据存储到数据库的相关内容，在 Scrapy 中可以实现 Item Pipeline 完成数据库存储的任务，我们以 SQLite、MySQL、MongoDB 和 Redis 几种常用数据库为例，先讲解在 Python 中将数据写入各种数据库的方法，然后实现相应的 Item Pipeline。

第 13 章

使用 HTTP 代理

大家可能都有过给浏览器设置 HTTP 代理的经验，HTTP 代理服务器可以比作客户端与 Web 服务器（网站）之间的一个信息中转站，客户端发送的 HTTP 请求和 Web 服务器返回的 HTTP 响应通过代理服务器转发给对方，如图 13-1 所示。

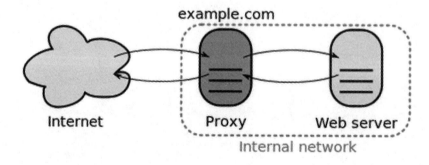

图 13-1

爬虫程序在爬取某些网站时也需要使用代理，例如：

- 由于网络环境因素，直接爬取速度太慢，使用代理提高爬取速度。
- 某些网站对用户的访问速度进行限制，爬取过快会被封禁 ip，使用代理防止被封禁。
- 由于地方法律或政治原因，某些网站无法直接访问，使用代理绕过访问限制。

这一章我们来学习 Scrapy 爬虫如何使用代理进行爬取。

13.1 HttpProxyMiddleware

Scrapy 内部提供了一个下载中间件 HttpProxyMiddleware，专门用于给 Scrapy 爬虫设置代理。

13.1.1 使用简介

HttpProxyMiddleware 默认便是启用的，它会在系统环境变量中搜索当前系统代理（名字格式为 xxx_proxy 的环境变量），作为 Scrapy 爬虫使用的代理。

假设我们现在有两台在云上搭建好的代理服务器：

http://116.29.35.201:8118
http://197.10.171.143:8118

为本机的 Scrapy 爬虫分别设置发送 HTTP 和 HTTPS 请求时所使用的代理，只需在 bash 中添加环境变量：

```
$ export http_proxy="http://116.29.35.201:8118"      # 为 HTTP 请求设置代理
$ export https_proxy="http://197.10.171.143:8118"    # 为 HTTPS 请求设置代理
```

配置完成后，Scrapy 爬虫将会使用上面指定的代理下载页面，我们可以通过以下实验进行验证。

利用网站 http://httpbin.org 提供的服务可以窥视我们所发送的 HTTP(S)请求，如请求源 IP 地址、请求头部、Cookie 信息等。图 13-2 展示了该网站各种服务的 API 地址。

访问 http(s)://httpbin.org/ip 将返回一个包含请求源 IP 地址信息的 json 串，在 scrapy shell 中访问该 url，查看请求源 IP 地址：

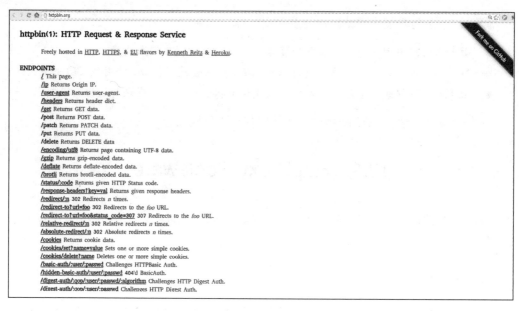

图 13-2

```
$ scrapy shell
...
>>> import json
>>> fetch(scrapy.Request('http://httpbin.org/ip'))      # 发送 HTTP 请求
[scrapy] DEBUG: Crawled (200)    (referer: None)
>>> json.loads(response.text)
{'origin': '116.29.35.201'}
>>> fetch(scrapy.Request('https://httpbin.org/ip'))     # 发送 HTTPS 请求
[scrapy] DEBUG: Crawled (200)    (referer: None)
>>> json.loads(response.text)
{'origin': '197.10.171.143'}
```

在上述实验中，分别以 HTTP 和 HTTPS 发送请求，使用 json 模块对响应结果进行解析，读取请求源 IP 地址（origin 字段），其值正是代理服务器的 IP。由此证明，Scrapy 爬虫使用了我们指定的代理。

上面我们使用的是无须身份验证的代理服务器，还有一些代理服务器需要用户提供账号、密码进行身份验证，验证成功后才提供代理服务，使用此类代理时，可按以下格式配置：

```
$ export http_proxy="http://liushuo:12345678@113.24.36.24:7777"
```

13.1.2 源码分析

虽然使用 HttpProxyMiddleware 很简单，但大家最好对其工作原理有所了解，以下是 HttpProxyMiddleware 的源码：

```python
import base64
from six.moves.urllib.request import getproxies, proxy_bypass
from six.moves.urllib.parse import unquote
try:
    from urllib2 import _parse_proxy
except ImportError:
    from urllib.request import _parse_proxy
from six.moves.urllib.parse import urlunparse

from scrapy.utils.httpobj import urlparse_cached
from scrapy.exceptions import NotConfigured
from scrapy.utils.python import to_bytes

class HttpProxyMiddleware(object):

    def __init__(self, auth_encoding='latin-1'):
        self.auth_encoding = auth_encoding
        self.proxies = {}
        for type, url in getproxies().items():
            self.proxies[type] = self._get_proxy(url, type)

        if not self.proxies:
            raise NotConfigured

    @classmethod
    def from_crawler(cls, crawler):
        auth_encoding = crawler.settings.get('HTTPPROXY_AUTH_ENCODING')
        return cls(auth_encoding)

    def _get_proxy(self, url, orig_type):
        proxy_type, user, password, hostport = _parse_proxy(url)
        proxy_url = urlunparse((proxy_type or orig_type, hostport, '', '', '', ''))

        if user:
            user_pass = to_bytes(
                '%s:%s' % (unquote(user), unquote(password)),
                encoding=self.auth_encoding)
```

```
            creds = base64.b64encode(user_pass).strip()
        else:
            creds = None

        return creds, proxy_url

    def process_request(self, request, spider):
        # ignore if proxy is already set
        if 'proxy' in request.meta:
            return

        parsed = urlparse_cached(request)
        scheme = parsed.scheme

        # 'no_proxy' is only supported by http schemes
        if scheme in ('http', 'https') and proxy_bypass(parsed.hostname):
            return

        if scheme in self.proxies:
            self._set_proxy(request, scheme)

    def _set_proxy(self, request, scheme):
        creds, proxy = self.proxies[scheme]
        request.meta['proxy'] = proxy
        if creds:
            request.headers['Proxy-Authorization'] = b'Basic ' + creds
```

分析代码如下：

- __init__ 方法

  在 HttpProxyMiddleware 的构造器中，使用 Python 标准库 urllib 中的 getproxies 函数在系统环境变量中搜索系统代理的相关配置（变量名格式为[协议]_proxy 的变量），调用 self._get_proxy 方法解析代理配置信息，并将其返回结果保存到 self.proxies 字典中，如果没有找到任何代理配置，就抛出 NotConfigured 异常，HttpProxyMiddleware 被弃用。

- _get_proxy 方法

  解析代理配置信息，返回身份验证信息（后面讲解）以及代理服务器 url。

- process_request 方法

  处理每一个待发送的请求，为没有设置过代理的请求（meta 属性不包含 proxy 字段的请求）调用 self._set_proxy 方法设置代理。

- _set_proxy 方法

  为一个请求设置代理，以请求的协议（HTTP 或 HTTPS）作为键，从代理服务器信息字典 self.proxies 中选择代理，赋给 request.meta 的 proxy 字段。对于需要身份验证的代理服务器，添加 HTTP 头部 Proxy-Authorization，其值是在_get_proxy 方法中计算得到的。

经分析得知，在 Scrapy 中为一个请求设置代理的本质就是将代理服务器的 url 填写到 request.meta['proxy']。

13.2 使用多个代理

利用 HttpProxyMiddleware 为爬虫设置代理时，对于一种协议（HTTP 或 HTTPS）的所有请求只能使用一个代理，如果想使用多个代理，可以在构造每一个 Request 对象时，通过 meta 参数的 proxy 字段手动设置代理：

```
request1 = Request('http://example.com/1', meta={'proxy': 'http://166.1.34.21:7117'})
request2 = Request('http://example.com/2', meta={'proxy': 'http://177.2.35.21:8118'})
request3 = Request('http://example.com/3', meta={'proxy': 'http://188.3.36.21:9119'})
```

按照与之前相同的做法，在 scrapy shell 进行实验，验证代理是否被使用：

```
$ scrapy shell
...
>>> from scrapy import Request
>>> req = Request('http://httpbin.org/ip', meta={'proxy': 'http://116.29.35.201:8118'})
>>> fetch(req)
[scrapy] DEBUG: Crawled (200)    (referer: None)
>>> json.loads(response.text)
{'origin': '116.29.35.201'}
>>> req = Request('https://httpbin.org/ip', meta={'proxy': 'http://197.10.171.143:8118'})
>>> fetch(req)
[scrapy] DEBUG: Crawled (200)    (referer: None)
>>> json.loads(response.text)
{'origin': '197.10.171.143'}
```

结果表明，Scrapy 爬虫同样使用了指定的代理服务器。

使用手动方式设置代理时，如果使用的代理需要身份验证，还需要通过 HTTP 头部的 Proxy-Authorization 字段传递包含用户账号和密码的身份验证信息。可以参考

HttpProxyMiddleware._get_proxy 中的相关实现，按以下过程生成身份验证信息：

（1）将账号、密码拼接成形如'user:passwd'的字符串 s1。
（2）按代理服务器要求对 s1 进行编码（如 utf8），生成 s2。
（3）再对 s2 进行 Base64 编码，生成 s3。
（4）将 s3 拼接到固定字节串 b'Basic '后面，得到最终的身份验证信息。

示例代码如下：

```
>>> from scrapy import Request
>>> import base64
>>> req = Request('http://httpbin.org/ip', meta={'proxy': 'http://116.29.35.201:8118'})
>>> user = 'liushuo'
>>> passwd = '12345678'
>>> user_passwd = ('%s:%s' % (user, passwd)).encode('utf8')
>>> user_passwd
b'liushuo:12345678'
>>> req.headers['Proxy-Authorization'] = b'Basic ' + base64.b64encode(user_passwd)
>>> fetch(req)
...
```

13.3 获取免费代理

现在，我们了解了如何为 Scrapy 爬虫设置代理，接下来的一个话题便是如何获取代理服务器。如果你感觉购买云服务器（亚马逊云或阿里云服务器等）自行搭建代理服务器的成本太高（但可靠、可控），那么可以通过 google 或 baidu 找到一些提供免费代理服务器信息的网站，例如：

http://proxy-list.org（国外）
https://free-proxy-list.net（国外）
http://www.xicidaili.com
http://www.proxy360.cn
http://www.kuaidaili.com

以 http://www.xicidaili.com 为例，图 13-3 所示为该网站"国内高匿代理"分类下的页面。

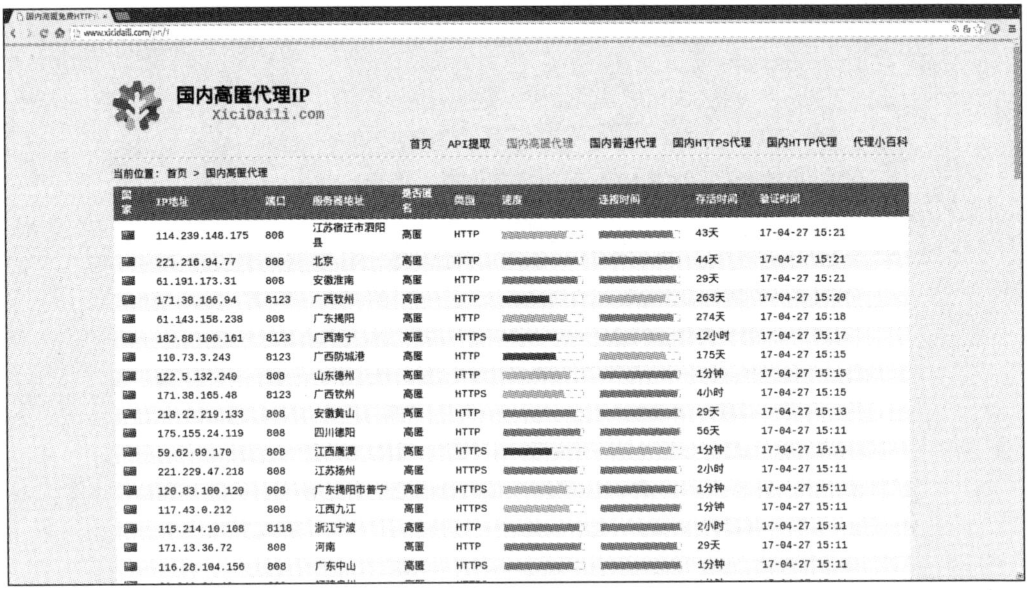

图 13-3

从中可以看出，该网站提供了大量的免费代理服务器信息，如果只需要少量的代理，从中选择几个就可以了，不过通常直觉告诉我们"多多益善"。接下来编写爬虫，爬取"国内高匿代理"分类下前 3 页的所有代理服务器信息，并验证每个代理是否可用。

创建 Scrapy 项目，取名为 proxy_example：

```
$ scrapy startproject proxy_example
```

该网站会检测用户发送的 HTTP 请求头部中的 User-Agent 字段，因此我们需要伪装成某种常规浏览器，在配置文件添加如下代码：

USER_AGENT = 'Mozilla/5.0 (X11; Linux x86_64) AppleWebKit/537.36 Chrome/41.0.2272.76'

实现 XiciSpider 爬取代理服务器信息，并过滤不可用代理，代码如下：

```
# -*- coding: utf-8 -*-
import scrapy
from scrapy import Request

import json

class XiciSpider(scrapy.Spider):
    name = "xici_proxy"
    allowed_domains = ["www.xicidaili.com"]
```

```python
def start_requests(self):
    # 爬取 http://www.xicidaili.com/nn/前 3 页
    for i in range(1, 4):
        yield Request('http://www.xicidaili.com/nn/%s' % i)

def parse(self, response):
    for sel in response.xpath('//table[@id="ip_list"]/tr[position()>1]'):
        # 提取代理的 IP、port、scheme(http or https)
        ip = sel.css('td:nth-child(2)::text').extract_first()
        port = sel.css('td:nth-child(3)::text').extract_first()
        scheme = sel.css('td:nth-child(6)::text').extract_first().lower()

        # 使用爬取到的代理再次发送请求到 http(s)://httpbin.org/ip，验证代理是否可用
        url = '%s://httpbin.org/ip' % scheme
        proxy = '%s://%s:%s' % (scheme, ip, port)

        meta = {
            'proxy': proxy,
            'dont_retry': True,
            'download_timeout': 10,

            # 以下两个字段是传递给 check_available 方法的信息，方便检测
            '_proxy_scheme': scheme,
            '_proxy_ip': ip,
        }

        yield Request(url, callback=self.check_available,
                      meta=meta, dont_filter=True)

def check_available(self, response):
    proxy_ip = response.meta['_proxy_ip']

    # 判断代理是否具有隐藏 IP 功能
    if proxy_ip == json.loads(response.text)['origin']:
        yield {
            'proxy_scheme': response.meta['_proxy_scheme'],
            'proxy': response.meta['proxy'],
        }
```

解释上述代码如下：

- 在 start_requests 方法中请求 http://www.xicidaili.com/nn 下的前 3 页，以 parse 方法作为页面解析函数。
- 在 parse 方法中提取一个页面中所有的代理服务器信息，这些代理未必都是可用的，因此使用爬取到的代理发送请求到 http(s)://httpbin.org/ip 验证其是否可用，以 check_available 方法作为页面解析函数。
- 能执行到 check_available 方法，意味着 response 对应请求所使用的代理是可用的。在 check_available 方法中，通过响应 json 串中的 origin 字段可以判断代理是否是匿名的（隐藏 ip），返回匿名代理。

运行爬虫，将可用的代理服务器保存到 json 文件中，供其他程序使用：

```
$ scrapy crawl xici_proxy -o proxy_list.json
...
$ cat proxy_list.json
[
{"proxy": "http://110.73.10.37:8123", "proxy_scheme": "http"},
{"proxy": "http://171.38.142.24:8123", "proxy_scheme": "http"},
{"proxy": "http://111.155.124.84:8123", "proxy_scheme": "http"},
{"proxy": "http://203.88.210.121:138", "proxy_scheme": "http"},
{"proxy": "http://182.88.191.195:8123", "proxy_scheme": "http"},
{"proxy": "http://121.31.151.231:8123", "proxy_scheme": "http"},
{"proxy": "http://203.93.0.115:80", "proxy_scheme": "http"},
{"proxy": "http://222.85.39.29:808", "proxy_scheme": "http"},
{"proxy": "http://175.155.25.26:808", "proxy_scheme": "http"},
{"proxy": "http://111.155.124.72:8123", "proxy_scheme": "http"},
{"proxy": "http://122.5.81.153:8118", "proxy_scheme": "http"},
{"proxy": "https://171.37.153.24:8123", "proxy_scheme": "https"},
{"proxy": "http://110.73.1.68:8123", "proxy_scheme": "http"},
{"proxy": "http://122.228.179.178:80", "proxy_scheme": "http"},
{"proxy": "http://121.31.151.226:8123", "proxy_scheme": "http"},
{"proxy": "http://171.38.171.168:8123", "proxy_scheme": "http"},
{"proxy": "http://218.22.219.133:808", "proxy_scheme": "http"},
{"proxy": "https://171.38.130.188:8123", "proxy_scheme": "https"},
{"proxy": "http://111.155.116.219:8123", "proxy_scheme": "http"},
{"proxy": "http://121.31.149.209:8123", "proxy_scheme": "http"},
{"proxy": "http://60.169.78.218:808", "proxy_scheme": "http"},
{"proxy": "http://171.38.158.227:8123", "proxy_scheme": "http"},
```

```
{"proxy": "http://121.31.150.224:8123", "proxy_scheme": "http"},
{"proxy": "http://111.155.124.78:8123", "proxy_scheme": "http"},
{"proxy": "http://182.90.83.104:8123", "proxy_scheme": "http"}
]
```

如结果所示，我们成功获取到了 20 多个可用的免费代理。

13.4 实现随机代理

本章开始部分曾提到，某些网站为防止爬虫爬取会对接收到的请求进行监测，如果在短时间内接收到了来自同一 IP 的大量请求，就判定该 IP 的主机在使用爬虫程序爬取网站，因而将该 IP 封禁（拒绝请求）。爬虫程序可以使用多个代理对此类网站进行爬取，此时单位时间的访问量会被多个代理分摊，从而避免封禁 IP。

下面我们基于 HttpProxyMiddleware 实现一个随机代理下载中间件。

在 middlewares.py 中实现 RandomHttpProxyMiddleware，代码如下：

```
from scrapy.downloadermiddlewares.httpproxy import HttpProxyMiddleware
from collections import defaultdict
import json
import random

class RandomHttpProxyMiddleware(HttpProxyMiddleware):

    def __init__(self, auth_encoding='latin-1', proxy_list_file=None):
        if not proxy_list_file:
            raise NotConfigured

        self.auth_encoding = auth_encoding
        # 分别用两个列表维护 HTTP 和 HTTPS 的代理，{'http': [...], 'https': [...]}
        self.proxies = defaultdict(list)

        # 从 json 文件中读取代理服务器信息，填入 self.proxies
        with open(proxy_list_file) as f:
            proxy_list = json.load(f)
            for proxy in proxy_list:
                scheme = proxy['proxy_scheme']
                url = proxy['proxy']
```

```
            self.proxies[scheme].append(self._get_proxy(url, scheme))

    @classmethod
    def from_crawler(cls, crawler):
        # 从配置文件中读取用户验证信息的编码
        auth_encoding = crawler.settings.get('HTTPPROXY_AUTH_ENCODING', 'latain-1')

        # 从配置文件中读取代理服务器列表文件（json）的路径
        proxy_list_file = crawler.settings.get('HTTPPROXY_PROXY_LIST_FILE')

        return cls(auth_encoding, proxy_list_file)

    def _set_proxy(self, request, scheme):
        # 随机选择一个代理
        creds, proxy = random.choice(self.proxies[scheme])
        request.meta['proxy'] = proxy
        if creds:
            request.headers['Proxy-Authorization'] = b'Basic ' + creds
```

解释上述代码如下：

- 仿照 HttpProxyMiddleware 构造器实现 RandomHttpProxyMiddleware 构造器，首先从代理服务器列表文件（配置文件中指定）中读取代理服务器信息，然后将它们按协议（HTTP 或 HTTPS）分别存入不同列表，由 self.proxis 字典维护。
- _set_proxy 方法负责为每一个 Request 请求设置代理，覆写 _set_proxy 方法（覆盖基类方法）。对于每一个 request，根据请求协议获取 self.proxis 中的代理服务器列表，然后从中随机抽取一个代理，赋值给 request.meta['proxy']。

在配置文件 settings.py 中启用 RandomHttpProxyMiddleware，并指定所要使用的代理服务器列表文件（json 文件），添加代码如下：

```
DOWNLOADER_MIDDLEWARES = {
    # 置于 HttpProxyMiddleware(750)之前
    'proxy_example.middlewares.RandomHttpProxyMiddleware': 745,
}

# 使用之前在 http://www.xicidaili.com/网站爬取到的代理
HTTPPROXY_PROXY_LIST_FILE='proxy_list.json'
```

最后编写一个 TestRandomProxySpider 测试该中间件，重复向 http(s)://httpbin.org/ip 发送请求，根据响应中的请求源 IP 地址信息判断代理使用情况：

```python
# -*- coding: utf-8 -*-
import scrapy
from scrapy import Request
import json

class TestRandomProxySpider(scrapy.Spider):
    name = "test_random_proxy"

    def start_requests(self):
        for _ in range(100):
            yield Request('http://httpbin.org/ip', dont_filter=True)
            yield Request('https://httpbin.org/ip', dont_filter=True)

    def parse(self, response):
        print(json.loads(response.text))
```

运行爬虫，观察输出：

```
$ scrapy crawl test_random_proxy
[scrapy] DEBUG: Crawled (200) <GET https://httpbin.org/ip> (referer: None)
{'origin': '171.38.130.188'}
[scrapy] DEBUG: Crawled (200) <GET http://httpbin.org/ip> (referer: None)
{'origin': '182.90.83.104'}
[scrapy] DEBUG: Crawled (200) <GET https://httpbin.org/ip> (referer: None)
{'origin': '171.38.130.188'}
[scrapy] DEBUG: Crawled (200) <GET http://httpbin.org/ip> (referer: None)
{'origin': '203.88.210.121'}
[scrapy] DEBUG: Crawled (200) <GET http://httpbin.org/ip> (referer: None)
{'origin': '203.88.210.121'}
...
```

结果表明，RandomHttpProxyMiddleware 工作良好，Scrapy 爬虫随机地使用了多个代理。

13.5 项目实战：爬取豆瓣电影信息

最后，我们来完成一个使用代理爬取的实战项目。豆瓣网的电影专栏是国内权威电影评分网站，其中包括海量影片信息，在浏览器中访问 https://movie.douban.com，并选择一个分类（如"豆瓣高分"），可看到如图 13-4 所示的影片列表页面。

图 13-4

单击其中一部影片，进入其页面（简称影片页面），如图 13-5 所示。

在影片页面中可以看到一部影片的基本信息，如导演、编剧、主演、类型等，我们可以编写爬虫在豆瓣电影中爬取大量影片信息。

图 13-5

13.5.1 项目需求

爬取豆瓣电影中"豆瓣高分"分类下的所有影片信息,需要爬取一部影片的信息字段如下:

- 导演
- 编剧
- 主演
- 类型

- 制片国家/地区
- 语言
- 上映日期
- 片长
- 又名

由于豆瓣网对爬取速度做了限制，高速爬取可能会被封禁 IP，因此使用代理进行爬取。

13.5.2 页面分析

首先分析影片列表页面，页面中的每一部电影都是通过 JavaScript 脚本加载的。单击页面最下方的"加载更多"，可以在 Chrome 开发者工具中捕获到 jQuery 发送的 HTTP 请求（加载更多影片），该请求返回了一个 json 串，如图 13-6 所示。

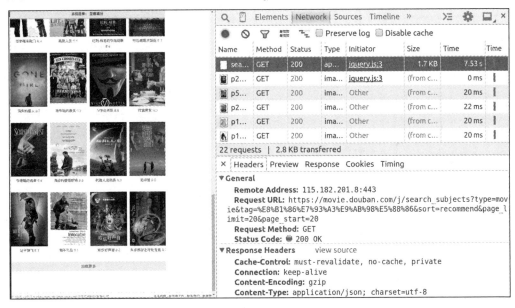

图 13-6

复制图中请求的 url，使用 scrapy shell 进行访问，查看其中 json 串的内容：

```
$ scrapy shell
'https://movie.douban.com/j/search_subjects?type=movie&tag=%E8%B1%86%E7%93%A3%E9%AB%98%E5%88%86&sort=recommend&page_limit=20&page_start=20'
...
```

```
>>> import json
>>> res = json.loads(response.body.decode('utf8'))
>>> res
{'subjects': [
  {'cover': 'https://img1.doubanio.com/view/movie_poster_cover/lpst/public/p511146957.jpg',
    'cover_x': 1538,
    'cover_y': 2159,
    'id': '1292001',
    'is_beetle_subject': False,
    'is_new': False,
    'playable': False,
    'rate': '9.2',
    'title': '海上钢琴师',
    'url': 'https://movie.douban.com/subject/1292001/'},
  {'cover': 'https://img1.doubanio.com/view/movie_poster_cover/lpst/public/p2360940399.jpg',
    'cover_x': 1500,
    'cover_y': 2145,
    'id': '25986180',
    'is_beetle_subject': False,
    'is_new': False,
    'playable': False,
    'rate': '8.2',
    'title': '釜山行',
    'url': 'https://movie.douban.com/subject/25986180/'},
  {'cover': 'https://img1.doubanio.com/view/movie_poster_cover/lpst/public/p2404978988.jpg',
    'cover_x': 703,
    'cover_y': 1000,
    'id': '26580232',
    'is_beetle_subject': False,
    'is_new': False,
    'playable': False,
    'rate': '8.7',
    'title': '看不见的客人',
    'url': 'https://movie.douban.com/subject/26580232/'},

...省略中间部分...

  {'cover': 'https://img3.doubanio.com/view/movie_poster_cover/lpst/public/p1454261925.jpg',
```

```
        'cover_x': 2181,
        'cover_y': 3120,
        'id': '6786002',
        'is_beetle_subject': False,
        'is_new': False,
        'playable': False,
        'rate': '9.1',
        'title': '触不可及',
        'url': 'https://movie.douban.com/subject/6786002/'},
       {'cover': 'https://img3.doubanio.com/view/movie_poster_cover/lpst/public/p2411622136.jpg',
        'cover_x': 1000,
        'cover_y': 1500,
        'id': '26354572',
        'is_beetle_subject': False,
        'is_new': False,
        'playable': True,
        'rate': '8.2',
        'title': '欢乐好声音',
        'url': 'https://movie.douban.com/subject/26354572/'},
       {'cover': 'https://img3.doubanio.com/view/movie_poster_cover/lpst/public/p1280323646.jpg',
        'cover_x': 1005,
        'cover_y': 1437,
        'id': '1299398',
        'is_beetle_subject': False,
        'is_new': False,
        'playable': False,
        'rate': '8.9',
        'title': '大话西游之月光宝盒',
        'url': 'https://movie.douban.com/subject/1299398/'}
]}
```

如上所示，返回结果（json）中的'subjects'字段是一个影片信息列表，一共有 20 项，每一项都是一部影片的信息，其中可以找到影片片名（title）、评分（rate）以及影片页面 url 等信息。

连续单击加载按钮，捕获更多 jQuery 发送的 HTTP 请求，可以总结出其 url 的规律：

- type 参数：类型，movie 代表电影。
- tag 参数：分类标签，当前为"豆瓣高分"。

- page_start：从第几部影片开始加载，即结果列表中第一部影片在服务器端的序号。
- page_limit：期望获取的影片信息的数量，当前为 20。

我们可以通过分析出的 API，每次获取固定数量的影片信息，从中提取每一个影片页面的 url，例如：

先获取 20 部影片信息：

[BASE_URL]?type=movie&tag=豆瓣高分&sort=recommend&page_limit=20&page_start=0

再获取 20 部影片信息：

[BASE_URL]?type=movie&tag=豆瓣高分&sort=recommend&page_limit=20&page_start=20

再获取 20 部影片信息：

[BASE_URL]?type=movie&tag=豆瓣高分&sort=recommend&page_limit=20&page_start=40

……

直到返回结果中的影片信息列表为空，说明没有影片了。

接下来分析影片页面。在 scrapy shell 中下载任意一个影片页面，并调用 view 函数在浏览器中查看页面，如图 13-7 所示。

图 13-7

从图 13-7 中可以看出，影片的信息在<div id="info">中，其中每一个信息字段名（"导演","编剧"等）都位于一个<span class="pl">中，比较容易提取，但字段的值很难找到统一的规律，我们可以使用 XPath 的 string 函数将<div id="info">中的所有文本提取到一个字符串，然后用提取到的字段名分割该字符串，得到其中值的部分。

首先提取包含所有信息的字符串：

```
>>> info = response.css('div#info').xpath('string(.)').extract_first()
>>> print(info)
导演: 涅提•蒂瓦里
编剧: 比于什•古普塔 / 施热亚•简
主演: 阿米尔•汗 / 法缇玛•萨那•纱卡 / 桑亚•玛荷塔 / 阿帕尔夏克提•库拉那 / 沙克希•坦沃
 / 泽伊拉•沃西姆 / 苏哈妮•巴特纳格尔 / 里特维克•萨霍里 / 吉里什•库卡尼',
类型 / 剧情 / 传记 / 运动
制片国家/地区: 印度
语言: 印地语
上映日期: 2017-05-05(中国大陆) / 2016-12-23(印度)
片长: 161 分钟(印度) / 140 分钟(中国大陆)
又名: 我和我的冠军女儿(台) / 打死不离 3 父女(港) / 摔跤吧！老爸 / 摔跤家族 / दंगल / Wrestling Competition
IMDb 链接: tt5074352
```

再提取所有字段名到一个列表：

```
>>> fields = [s.strip().replace(':', '') for s in \
...           response.css('div#info span.pl::text').extract()]
>>> fields
['导演','编剧','主演','类型','制片国家/地区','语言','上映日期','片长','又名','IMDb 链接']
```

使用字段名对 info 进行分割，得到所有值的列表：

```
>>> import re
>>> values = [re.sub('\s+', ' ', s.strip()) for s in \
...           re.split('\s*(?:%s):\s*' % '|'.join(fields), info)][1:]
>>> values
['涅提•蒂瓦里',
 '比于什•古普塔 / 施热亚•简',
 '阿米尔•汗 / 法缇玛•萨那•纱卡 / 桑亚•玛荷塔 / 阿帕尔夏克提•库拉那 / 沙克希•坦沃 / 泽伊拉•沃西姆 / 苏哈妮•巴特纳格尔 / 里特维克•萨霍里 / 吉里什•库卡尼',
```

```
'剧情 / 传记 / 运动',
'印度',
'印地语',
'2017-05-05(中国大陆) / 2016-12-23(印度)',
'161 分钟(印度) / 140 分钟(中国大陆)',
'我和我的冠军女儿(台) / 打死不离 3 父女(港) / 摔跤吧！老爸 / 摔跤家族 / दंगल /
Wrestling Competition',
'tt5074352']
```

最后，使用以上两个列表构造影片信息字典：

```
>>> dict(zip(fields, values))
{'IMDb 链接': 'tt5074352',
 '上映日期': '2017-05-05(中国大陆) / 2016-12-23(印度)',
 '主演': '阿米尔•汗 / 法缇玛•萨那•纱卡 / 桑亚•玛荷塔 / 阿帕尔夏克提•库拉那 / 沙克希•坦沃 /
泽伊拉•沃西姆 / 苏哈妮•巴特纳格尔 / 里特维克•萨霍里 / 吉里什•库卡尼',
 '制片国家/地区': '印度',
 '又名': '我和我的冠军女儿(台) / 打死不离 3 父女(港) / 摔跤吧！老爸 / 摔跤家族 / दंगल /
Wrestling Competition',
 '导演': '涅提•蒂瓦里',
 '片长': '161 分钟(印度) / 140 分钟(中国大陆)',
 '类型': '剧情 / 传记 / 运动',
 '编剧': '比于什•古普塔 / 施热亚•简',
 '语言': '印地语'}
```

经上述操作，我们得到了除片名和评分之外的所有信息，片名和评分信息可以在 json 串中获取。

到此，页面分析的工作完成了。

13.5.3 编码实现

创建 Scrapy 项目，取名为 douban_movie：

```
$ scrapy startproject douban_movie
```

在页面分析中，我们详细阐述了爬取过程，现在可以轻松实现 MoviesSpider 了，代码如下：

```
# -*- coding: utf-8 -*-
import scrapy
```

```python
from scrapy import Request
import json
import re
from pprint import pprint

class MoviesSpider(scrapy.Spider):
    BASE_URL = 'https://movie.douban.com/j/search_subjects?type=movie&tag=%s&sort=recommend&page_limit=%s&page_start=%s'
    MOVIE_TAG = '豆瓣高分'
    PAGE_LIMIT = 20
    page_start = 0

    name = "movies"
    start_urls = [BASE_URL % (MOVIE_TAG, PAGE_LIMIT, page_start)]

    def parse(self, response):
        # 使用 json 模块解析响应结果
        infos = json.loads(response.body.decode('utf-8'))

        # 迭代影片信息列表
        for movie_info in infos['subjects']:
            movie_item = {}

            # 提取"片名"和"评分", 填入 item.
            movie_item['片名'] = movie_info['title']
            movie_item['评分'] = movie_info['rate']

            # 提取影片页面 url, 构造 Request 发送请求, 并将 item 通过 meta 参数传递给影片页面解析函数
            yield Request(movie_info['url'], callback=self.parse_movie,
                          meta={'_movie_item': movie_item})

        # 如果 json 结果中包含的影片数量小于请求数量, 说明没有影片了, 否则继续搜索
        if len(infos['subjects']) == self.PAGE_LIMIT:
            self.page_start += self.PAGE_LIMIT
            url = self.BASE_URL % (self.MOVIE_TAG, self.PAGE_LIMIT, self.page_start)
            yield Request(url)
```

```python
def parse_movie(self, response):
    # 从 meta 中提取已包含"片名"和"评分"信息的 item
    movie_item = response.meta['_movie_item']

    # 获取整个信息字符串
    info = response.css('div.subject div#info').xpath('string(.)').extract_first()

    # 提取所有字段名
    fields= [s.strip().replace(':', '') for s in \
                response.css('div#info span.pl::text').extract()]

    # 提取所有字段的值
    values = [re.sub('\s+', '', s.strip()) for s in \
                re.split('\s*(?:%s):\s*' % '|'.join(fields), info)][1:]

    # 将所有信息填入 item
    movie_item.update(dict(zip(fields, values)))

    yield movie_item
```

为了使用代理进行爬取，我们将之前实现的 RandomHttpProxyMiddleware 复制到该项目中。

在配置文件 settings.py 中添加如下配置：

```python
# 我们的爬取不符合豆瓣爬取规则，强制爬取
ROBOTSTXT_OBEY = False

# 伪装成常规浏览器
USER_AGENT = 'Mozilla/5.0 (X11; Linux x86_64) Chrome/42.0.2311.90 Safari/537.36'

# 可选。设置下载延迟，防止代理被封禁 IP，依据代理数量而定
DOWNLOAD_DELAY = 0.5

# 启用随机代理中间件
DOWNLOADER_MIDDLEWARES = {
    'douban_movie.middlewares.RandomHttpProxyMiddleware': 745,
}
# 指定所要使用的代理服务器列表文件
HTTPPROXY_PROXY_LIST_FILE = 'proxy_list.json'
```

为了爬取稳定，使用在云服务器上自行搭建的代理服务器，手工编辑代理服务器列表文件 proxy_list.json：

```
[
{"proxy_scheme": "https", "proxy": "http://116.29.35.201:8118"},
{"proxy_scheme": "https", "proxy": "http://197.10.171.143:8118"},
{"proxy_scheme": "https", "proxy": "http://112.78.43.67:8118"},
{"proxy_scheme": "https", "proxy": "http://124.59.42.145:8118"}
]
```

最后，运行爬虫，将结果保存到文件 moveis.json：

```
$ scrapy crawl movies -o movies.json
```

在 Python 中观察爬取结果，代码如下：

```
>>> import json
>>> movies = json.load(open('movies.json'))
>>> for movie in movies:
...     print(movie['片名'], movie['评分'], movie['导演'])
...
忠犬八公的故事 9.2 拉斯·霍尔斯道姆
楚门的世界 9.0 彼得·威尔
怦然心动 8.9 罗伯·莱纳
泰坦尼克号 9.2 詹姆斯·卡梅隆
血战钢锯岭 8.7 梅尔·吉布森
驴得水 8.3 周申 / 刘露
星际穿越 9.1 克里斯托弗·诺兰
盗梦空间 9.2 克里斯托弗·诺兰
阿甘正传 9.4 罗伯特·泽米吉斯
千与千寻 9.2 宫崎骏
三傻大闹宝莱坞 9.1 拉吉库马尔·希拉尼
霸王别姬 9.5 陈凯歌
你的名字 8.5 新海诚
金刚狼 3：殊死一战 8.3 詹姆斯·曼高德
疯狂动物城 9.2 拜伦·霍华德 / 瑞奇·摩尔 / 杰拉德·布什
爱乐之城 8.3 达米恩·查泽雷
大话西游之月光宝盒 8.9 刘镇伟
触不可及 9.1 奥利维·那卡什 / 艾力克·托兰达
无间道 9.0 刘伟强 / 麦兆辉
```

让子弹飞 8.7 姜文

...省略中间部分...

哈尔的移动城堡 8.8 宫崎骏
阿凡达 8.6 詹姆斯·卡梅隆
教父 9.2 弗朗西斯·福特·科波拉
罗马假日 8.9 威廉·惠勒
龙猫 9.1 宫崎骏
火星救援 8.4 雷德利·斯科特
超能陆战队 8.6 唐·霍尔／克里斯·威廉姆斯
欢乐好声音 8.2 加斯·詹宁斯／克里斯托夫·卢尔德莱
少年派的奇幻漂流 9.0 李安
釜山行 8.2 延尚昊
大话西游之大圣娶亲 9.2 刘镇伟
这个杀手不太冷 9.4 吕克·贝松
海上钢琴师 9.2 朱塞佩·托纳多雷
宣告黎明的露之歌 8.1 汤浅政明
肖申克的救赎 9.6 弗兰克·德拉邦特
摔跤吧！爸爸 9.2 涅提·蒂瓦里
>>> len(movies)
500

如上所示，我们成功爬取了"豆瓣高分"分类下的 500 部影片信息。

13.6 本章小结

本章学习了 Scrapy 爬虫如何使用代理进行爬取，首先介绍了两种设置代理的方法：

- 使用下载中间件 HttpProxyMiddleware（自动）。
- 在构造 Request 对象时通过 meta 参数设置（手动）。

前者使用简单，只需通过环境变量配置即可；后者可在某些特殊场景下使用。我们以如何使用多个代理进行了举例，随后讲解了在网络上获取免费代理的方法，并利用获取的免费代理实现了一个随机代理中间件。最后，运用本章所学的知识完成一个实战项目，使用代理爬取了豆瓣网中的电影信息。

第 14 章 分布式爬取

由于受到计算能力和网络带宽的限制，单台计算机上运行的爬虫在爬取的数据量较大时，需要耗费很长的时间。分布式爬取的思想是"人多力量大"，在网络中的多台计算机上同时运行爬虫程序，共同完成一个大型爬取任务。这一章来学习使用 Scrapy 框架进行分布式爬取。

Scrapy 本身并不是一个为分布式爬取而设计的框架，但第三方库 scrapy-redis 为其拓展了分布式爬取的功能，两者结合便是一个分布式 Scrapy 爬虫框架。在分布式爬虫框架中，需要使用某种通信机制协调各个爬虫的工作，让每一个爬虫明确自己的任务，其中包括：

（1）当前的爬取任务，即下载+提取数据（分配任务）。
（2）当前爬取任务是否已经被其他爬虫执行过（任务去重）。
（3）如何存储爬取到的数据（数据存储）。

scrapy-redis 利用 Redis 数据库作为多个爬虫的数据共享实现以上功能，接下来我们学习如何使用 scrapy-redis 进行分布式爬取。

14.1　Redis 的使用

首先来学习 Redis 数据库的使用。Redis 是一个速度非常快的非关系型数据库，使用内存作为主存储，内存中的数据也可以被持久化到硬盘。Redis 以键值形式（key-value）存储数据，其中的值可以分为以下 5 种类型：

- 字符串（string）
- 哈希（hash）
- 列表（list）
- 集合（set）
- 有序集合（zset）

14.1.1　安装 Redis

接下来安装 Redis，在 Ubuntu 下可以使用 apt-get 安装：

```
sudo apt-get install redis-server
```

Redis 数据库进程是一个网络服务器，可以使用以下命令开启/重启/停止 Redis：

```
$ sudo service redis-server start          # 开启 Redis
$ sudo service redis-server restart        # 重启 Redis
$ sudo service redis-server start          # 停止 Redis
```

默认情况下，Redis 会在 127.0.0.1:6379 上开启服务，可以使用 netstat 命令进行查询：

```
$ netstat -ntl
Proto Recv-Q Send-Q Local Address           Foreign Address         State
...
tcp        0      0 127.0.0.1:6379          *:*                     LISTEN
...
```

此时，我们运行的 Redis 仅能被本机使用，因为它只接收来自本机（localhost）的请求。如果想让 Redis 能被其他计算机访问，需要在配置文件中修改 Redis 服务器的监听地址，在 Redis 配置文件/etc/redis/redis.conf 中做以下修改：

```
$ sudo vi /etc/redis/redis.conf
...
# bind 127.0.0.1
```

```
bind 0.0.0.0              # 接收来自任意 IP 的请求
...
```

重启 Redis 后，修改生效：

```
$ sudo service redis-server restart
Stopping redis-server: redis-server.
Starting redis-server: redis-server.
$ netstat -ntl
Proto Recv-Q Send-Q Local Address        Foreign Address      State
...
$ tcp     0      0 0.0.0.0:6379          0.0.0.0:*            LISTEN
...
```

现在，可以在任意机器上使用客户端连接 Redis 数据库：

```
$ redis-cli -h 192.168.0.103       # 使用-h 参数指定主机 IP
192.168.1.102:6379> PING           # 测试链接数据库是否成功
PONG
```

14.1.2 Redis 基本命令

由于篇幅有限，这里仅介绍一些 Redis 基本使用命令，按照值的 5 种类型依次讲解。想学习 Redis 更多详细内容，请参考相关书籍或 Redis 官方网站 https://redis.io。

字符串

Redis 的字符串（String）可以存储字符串、整数、浮点数（数字也是字符串）。String 命令及描述如表 14-1 所示。

表 14-1 String 命令及描述

String 命令	描述
SET key value	设置字符串 key 的值
GET key	获取字符串 key 的值
DEL key	删除 key

```
192.168.1.102:6379> SET x abcde
OK
192.168.1.102:6379> GET x
"abcde"
```

```
192.168.1.102:6379> DEL x
(integer) 1
192.168.1.102:6379> GET x
(nil)
```

列表

Redis 的列表（List）可以有序地存储多个字符串。List 命令及描述如表 14-2 所示。

表 14-2 List命令及描述

List 命令	描述
LPUSH key value1 [value2...]	在列表 key 左端插入一个或多个值
RPUSH key value1 [value2...]	在列表 key 右端插入一个或多个值
LPOP key	从列表 key 左端弹出一个值
RPOP key	从列表 key 右端弹出一个值
LINDEX key index	获取列表 key 中 index 位置的值
LRANGE key start end	获取列表 key 中位置在[start, end]范围的值
LLEN key	获取列表 key 的长度

```
192.168.1.102:6379> RPUSH color orange          # 右边入队 1 个值
(integer) 1
192.168.1.102:6379> RPUSH color red yellow blue # 右边入队 3 个值
(integer) 4
192.168.1.102:6379> LINDEX color 2              # 获取位置 2 的值
"yellow"
192.168.1.102:6379> LRANGE color 0 3            # 获取位置在 0~3 范围的值
1) "orange"
2) "red"
3) "yellow"
4) "blue"
192.168.1.102:6379> LPOP color                  # 左边出队 1 个值
"orange"
192.168.1.102:6379> LRANGE color 0 -1           # 获取全部值（负引索表示倒数第几个）
1) "red"
2) "yellow"
3) "blue"
```

哈希

Redis 的哈希（Hash）可以存储多个键值对，其中的键和值都是字符串。Hash 命令及描述如表 14-3 所示。

表 14-3　Hash 命令与描述

Hash 命令	描　　述
HSET key field value	将哈希 key 的 field 字段赋值为 value
HDEL key field1 [field2...]	删除哈希 key 的一个或多个字段
HGET key field	获取哈希 key 的 field 字段的值
HGETALL key	获取哈希 key 的所有字段和值

```
192.168.1.102:6379> HSET point x 34            # 赋值 x、y、z 三个字段
(integer) 1
192.168.1.102:6379> HSET point y 55
(integer) 1
192.168.1.102:6379> HSET point z 47
(integer) 1
192.168.1.102:6379> HGET point x               # 获取字段 x 的值
"34"
192.168.1.102:6379> HGETALL point              # 获取所有字段和值
1) "x"
2) "34"
3) "y"
4) "55"
5) "z"
6) "47"
192.168.1.102:6379> HDEL point y               # 删除字段 y
(integer) 1
192.168.1.102:6379> HGETALL point
1) "x"
2) "34"
3) "z"
4) "47"
```

集合

Redis 中的集合（Set）可以存储多个唯一的字符串。Set 命令及描述如表 14-4 所示。

表 14-4 Set 命令及描述

Set 命令	描述
SADD key member1 [member2...]	向集合 key 中添加一个或多个成员
SREM key member1 [member2...]	删除集合 key 中一个或多个成员
SMEMBERS key	获取集合 key 中所有成员
SCARD key	获取集合 key 中成员数量
SISMEMBER key member	判断 member 是否是集合 key 的成员

```
192.168.1.102:6379> SADD color_set white black red blue    # 添加 4 个成员
(integer) 4
192.168.1.102:6379> SMEMBERS color_set                     # 获取全部成员
1) "blue"
2) "red"
3) "black"
4) "white"
192.168.1.102:6379> SADD color_set white                   # 添加已存在的成员会失败
(integer) 0
192.168.1.102:6379> SCARD color_set                        # 查询成员个数
(integer) 4
192.168.1.102:6379> SREM color_set red                     # 删除 1 个成员
(integer) 1
192.168.1.102:6379> SMEMBERS color_set
1) "black"
2) "white"
3) "blue"
192.168.1.102:6379> SISMEMBER color_set blue               # 判断某成员是否存在
(integer) 1
192.168.1.102:6379> SISMEMBER color_set red
(integer) 0
```

有序集合

Redis 中的有序集合（ZSet）与集合（Set）类似，可以存储多个唯一的字符串，但在有序集合中，每个成员都有一个分数，所有成员按给定分数在集合中有序排列。Zset 命令及描述如表 14-5 所示。

表 14-5　ZSet 命令及描述

ZSet 命令	描　　述
ZADD key score1 member1 [score2 member2...]	向有序集合 key 中添加一个或多个成员
ZREM key member [member2 ...]	删除有序集合 key 中一个或多个成员
ZRANGE key start stop	获取有序集合 key 中位置在[start, stop]范围的所有成员
ZRANGEBYSCORE key min max	获取有序集合 key 中分值在[start, stop]范围的所有成员

```
192.168.1.102:6379> ZADD language 3 python 6 perl 9 lisp    # 添加 3 个成员
(integer) 3
192.168.1.102:6379> ZADD language 1 C 4 java 5 C++          # 再添加 3 个成员
(integer) 3
192.168.1.102:6379> ZREM language C++                       # 删除 1 个成员
(integer) 1
192.168.1.102:6379> ZRANGE language 2 4                     # 获取位置在 2~4 范围的成员
1) "java"
2) "perl"
3) "lisp"
192.168.1.102:6379> ZRANGEBYSCORE language 3 6              # 获取分值在 3~6 范围的成员
1) "python"
2) "java"
3) "perl"
```

14.1.3　Python 访问 Redis

Redis 支持多种语言 API，在 Python 中可以使用第三方库 redis-py 访问 Redis 数据库。使用 pip 安装 redis-py：

```
$ sudo pip install redis
```

redis-py 的使用非常简单，只需先连接 Redis 获得连接对象，然后调用该对象上与每一条 Redis 命令相对应的方法即可。请看下面的简单示例：

```
>>> import redis
>>> r = redis.StrictRedis(host='localhost', port=6379)      # 连接 Redis，返回连接对象
>>> r.set('s', 'hello world')                               # 字符串
True
```

```
>>> r.get('s')
b'hello world'
>>> r.rpush('queue', 1, 2, 3)                    # 列表
3
>>> r.lpop('queue')
b'1'
>>> r.llen('queue')
2
>>> r.lrange('queue', 0, -1)
[b'2', b'3']
```

Redis 简介部分到此结束了,希望这部分内容能够帮助大家读懂 scrapy-redis 的源码。

14.2 scrapy-redis 源码分析

在使用 scrapy-redis 前,我们先来分析 scrapy-redis 的源码,了解其实现原理。

使用 git 从 github 网站下载 scrapy-redis 源码:

```
$ git clone https://github.com/rolando/scrapy-redis
$ cd scrapy-redis/src/
$ tree
.
└── scrapy_redis
    ├── connection.py
    ├── defaults.py
    ├── dupefilter.py
    ├── __init__.py
    ├── picklecompat.py
    ├── pipelines.py
    ├── queue.py
    ├── scheduler.py
    ├── spiders.py
    └── utils.py

1 directory, 10 files
```

scrapy-redis 的源码并不多,因为它仅是利用 Redis 数据库重新实现了 Scrapy 中的某些组件。

对于一个分布式爬虫框架，需要解决以下两个最基本的问题。

- 分配爬取任务：为每个爬虫分配不重复的爬取任务。
- 汇总爬取数据：将所有爬虫爬取到的数据汇总到一处。

接下来我们看 scrapy-redis 是如何解决的。

14.2.1 分配爬取任务部分

scrapy-redis 为多个爬虫分配爬取任务的方式是：让所有爬虫共享一个存在于 Redis 数据库中的请求队列（替代各爬虫独立的请求队列），每个爬虫从请求队列中获取请求，下载并解析页面后，将解析出的新请求再添加到请求队列中，因此每个爬虫既是下载任务的生产者又是消费者。

为实现多个爬虫的任务分配，scrapy-redis 重新实现了以下组件：

- 基于 Redis 的请求队列（优先队列、FIFO、LIFO）。
- 基于 Redis 的请求去重过滤器（过滤掉重复的请求）。
- 基于以上两个组件的调度器。

1. 调度器的实现

首先来看调度器 Scheduler 的实现，它位于 scheduler.py 中，Scheduler 的核心代码如下：

```
import importlib
import six

from scrapy.utils.misc import load_object
from . import connection, defaults


class Scheduler(object):
    ...
    def open(self, spider):
        ...
        # 初始化请求队列
        try:
            self.queue = load_object(self.queue_cls)(
                server=self.server,
                spider=spider,
                key=self.queue_key % {'spider': spider.name},
```

```python
            serializer=self.serializer,
        )
    except TypeError as e:
        raise ValueError("Failed to instantiate queue class '%s': %s",
                         self.queue_cls, e)



    # 初始化去重过滤器
    try:
        self.df = load_object(self.dupefilter_cls)(
            server=self.server,
            key=self.dupefilter_key % {'spider': spider.name},
            debug=spider.settings.getbool('DUPEFILTER_DEBUG'),
        )
    except TypeError as e:
        raise ValueError("Failed to instantiate dupefilter class '%s': %s",
                         self.dupefilter_cls, e)
    ...

...
def enqueue_request(self, request):
    # 调用去重过滤器的 request_seen 方法,判断该 request 对应的页面是否已经爬取过
    # 如果页面已经爬取过,且用户没有强制忽略过滤,就直接返回 False
    if not request.dont_filter and self.df.request_seen(request):
        self.df.log(request, self.spider)
        return False
    if self.stats:
        self.stats.inc_value('scheduler/enqueued/redis', spider=self.spider)
    # 将 request 入队到请求队列
    self.queue.push(request)
    return True

def next_request(self):
    block_pop_timeout = self.idle_before_close
    # 从请求队列出队一个 request
    request = self.queue.pop(block_pop_timeout)
    if request and self.stats:
        self.stats.inc_value('scheduler/dequeued/redis', spider=self.spider)
```

```
        return request
    ...
```

分析上述代码如下：

- 调度器中最核心的两个方法是 enqueue_request 和 next_request，它们分别对应请求的入队和出队操作。Spider 提交的 Request 对象最终由 Scrapy 引擎调用 enqueue_request 添加到请求队列中，Scrapy 引擎同时也调用 next_request 从请求队列中取出请求，送给下载器下载。
- self.queue 和 self.df 分别是请求队列和去重过滤器对象。在 enqueue_request 方法中，使用去重过滤器的 request_seen 方法判断 request 是否重复，即 request 对应的页面是否已经爬取过，如果用户没有强制忽略过滤，并且 request 是重复的，就抛弃该 request，并直接返回 False，否则调用 self.queue 的 push 方法将 request 入队，返回 True。在 next_request 方法中，调用 self.queue 的 pop 方法出队一个 request 并返回。

再来看一下创建请求队列和去重过滤器对象的相关代码：

```
import importlib
import six

from scrapy.utils.misc import load_object
from . import connection, defaults

class Scheduler(object):
    def __init__(self, server, ...,
                 queue_cls=defaults.SCHEDULER_QUEUE_CLASS,
                 dupefilter_cls=defaults.SCHEDULER_DUPEFILTER_CLASS):
        ...
        self.server = server
        self.queue_cls = queue_cls
        self.dupefilter_cls = dupefilter_cls
        ...

    @classmethod
    def from_settings(cls, settings):
        ...
        # 从配置中读取请求队列和去重过滤器的类名
        optional = {
            ...
```

```
            'queue_cls': 'SCHEDULER_QUEUE_CLASS',
            'dupefilter_cls': 'DUPEFILTER_CLASS',
            ...
        }
        for name, setting_name in optional.items():
            val = settings.get(setting_name)
            if val:
                kwargs[name] = val
        ...
        server = connection.from_settings(settings)
        server.ping()

        return cls(server=server, **kwargs)
    ...

    def open(self, spider):
        ...
        # 初始化请求队列
        try:
            self.queue = load_object(self.queue_cls)(
                server=self.server,
                spider=spider,
                key=self.queue_key % {'spider': spider.name},
                serializer=self.serializer,
            )
        except TypeError as e:
            raise ValueError("Failed to instantiate queue class '%s': %s",
                             self.queue_cls, e)

        # 初始化去重过滤器
        try:
            self.df = load_object(self.dupefilter_cls)(
                server=self.server,
                key=self.dupefilter_key % {'spider': spider.name},
                debug=spider.settings.getbool('DUPEFILTER_DEBUG'),
            )
        except TypeError as e:
            raise ValueError("Failed to instantiate dupefilter class '%s': %s",
```

```
                    self.dupefilter_cls, e)
            ...
        ...
```

self.queue 和 self.df 的创建是在 open 方法中调用 load_object 方法完成的，load_object 方法的参数是类的导入路径（如 scrapy_redis.queue.PriorityQueue），这种实现的好处是用户可以使用字符串在配置文件中灵活指定想要使用的队列类和过滤器类。self.queue_cls 和 self.dupefilter_cls 便是从配置文件中读取的导入路径（或默认值）。

2. 请求队列的实现

接下来看基于 Redis 的请求队列的实现。在 queque.py 中，包含以下 3 种请求队列：

- PriorityQueue 优先级队列（默认）
- FifoQueue 先进先出队列
- LifoQueue 后进先出队列

我们分析其中代码简短且容易理解的 FifoQueue，代码如下：

```python
class FifoQueue(Base):
    """Per-spider FIFO queue"""

    def __len__(self):
        """Return the length of the queue"""
        return self.server.llen(self.key)

    def push(self, request):
        """Push a request"""
        self.server.lpush(self.key, self._encode_request(request))

    def pop(self, timeout=0):
        """Pop a request"""
        if timeout > 0:
            # brpop: 阻塞的 rpop，可以设置超时
            data = self.server.brpop(self.key, timeout)
            if isinstance(data, tuple):
                data = data[1]
        else:
            data = self.server.rpop(self.key)
        if data:
            return self._decode_request(data)
```

- self.server 是 Redis 数据库的连接对象（可理解为 self.server = redis.StrictRedis(...)），该连接对象是在 Scheduler 的 from_settings 方法中创建的，在创建请求队列对象时，被传递给请求队列类的构造器。
- 观察在 self.server 上调用的方法可知，FifoQueue 使用 Redis 中的一个列表实现队列，该列表在数据库中的键为 self.key 的值，可以通过配置文件设置（SCHEDULER_QUEUE_KEY），默认为<spider_name>:requests。
- push 方法对应请求的入队操作，先调用基类的_encode_request 方法对 request 进行编码，然后调用 Redis 的 lpush 命令将其插入数据库中列表的最左端（入队）。
- pop 方法对应请求的出队操作，调用 Redis 的 rpop 或 brpop 命令从数据库中列表的最右端弹出一个经过编码的 request（出队），再调用基类的_decode_request 方法对其进行解码，然后返回。
- __len__ 方法调用 Redis 的 llen 命令获取数据库中列表的长度，即请求队列的长度。

下面是 FifoQueue、LifoQueue、PriorityQueue 共同基类 Base 的部分代码：

```
from . import picklecompat

class Base(object):
    """Per-spider base queue class"""

    def __init__(self, server, spider, key, serializer=None):
        """Initialize per-spider redis queue.

        Parameters
        ----------
        server : StrictRedis
            Redis client instance.
        spider : Spider
            Scrapy spider instance.
        key: str
            Redis key where to put and get messages.
        serializer : object
            Serializer object with ``loads`` and ``dumps`` methods.
        """

        if serializer is None:
            serializer = picklecompat
        ...
```

```
        self.server = server
        self.spider = spider
        self.key = key % {'spider': spider.name}
        self.serializer = serializer

    def _encode_request(self, request):
        """Encode a request object"""
        obj = request_to_dict(request, self.spider)
        return self.serializer.dumps(obj)

    def _decode_request(self, encoded_request):
        """Decode an request previously encoded"""
        obj = self.serializer.loads(encoded_request)
        return request_from_dict(obj, self.spider)
    ...
```

可以看到，在对 request 进行编、解码时，调用的是 self.serializer 的 dumps 和 loads 方法。self.serializer 同样可以通过配置文件指定（SCHEDULER_SERIALIZER），默认为 Python 标准库中的 pickle 模块。

3. 去重过滤器的实现

最后来看基于 Redis 的去重过滤器 RFPDupeFilter 的实现，它位于 dupefilter.py 中，部分代码如下：

```
...
from scrapy.dupefilters import BaseDupeFilter
from scrapy.utils.request import request_fingerprint
...
class RFPDupeFilter(BaseDupeFilter):
    ...
    def __init__(self, server, key, debug=False):
        """Initialize the duplicates filter.

        Parameters
        ----------
        server : redis.StrictRedis
            The redis server instance.
        key : str
```

```
                Redis key Where to store fingerprints.
        debug : bool, optional
                Whether to log filtered requests.
        """
        self.server = server
        self.key = key
        self.debug = debug
        self.logdupes = True
    ...
    def request_seen(self, request):
        fp = self.request_fingerprint(request)
        # This returns the number of values added, zero if already exists.
        added = self.server.sadd(self.key, fp)
        return added == 0

    def request_fingerprint(self, request):
        return request_fingerprint(request)
    ...
```

- self.server 是 Redis 数据库的连接对象，与 FifoQueue 中相同。
- 观察在 self.server 上调用的方法可知，RFPDupeFilter 使用 Redis 中的一个集合对请求进行去重，该集合在数据库中的键为 self.key 的值，可以通过配置文件设置（SCHEDULER_DUPEFILTER_KEY），默认为<spider_name>:dupefilter。
- request_fingerprint 方法用来获取一个请求的指纹，即请求的唯一标识，请求的指纹是使用 Python 标准库 hashlib 中的 sha1 算法计算得到的（详见 scrapy.utils.request 中的 request_fingerprint 函数）。
- request_seen 方法用来判断一个请求是否是重复的，先调用 request_fingerprint 方法计算 request 的指纹，然后调用 Redis 的 sadd 命令尝试将指纹添加到数据库中的集合中，根据 sadd 返回值判断请求是否重复，返回相应的布尔值结果，即重复返回 True，否则返回 False。

14.2.2 汇总爬取数据部分

1. RedisPipeline 的实现

在分布式爬虫框架中，各个主机爬取到的数据最终要汇总到一处，通常是某种数据库。scrapy-redis 提供了一个 Item Pipeline(RedisPipeline)，用于将各个爬虫爬取到的数据存入同一个 Redis 数据库中。

RedisPipeline 位于 pipeline.py 中，代码如下：

```
from scrapy.utils.misc import load_object
from scrapy.utils.serialize import ScrapyJSONEncoder
from twisted.internet.threads import deferToThread

from . import connection, defaults


default_serialize = ScrapyJSONEncoder().encode


class RedisPipeline(object):
    """Pushes serialized item into a redis list/queue

    Settings
    --------
    REDIS_ITEMS_KEY : str
        Redis key where to store items.
    REDIS_ITEMS_SERIALIZER : str
        Object path to serializer function.

    """

    def __init__(self, server,
                 key=defaults.PIPELINE_KEY,
                 serialize_func=default_serialize):
        """Initialize pipeline.

        Parameters
        ----------
        server : StrictRedis
            Redis client instance.
        key : str
            Redis key where to store items.
        serialize_func : callable
            Items serializer function.

        """
        self.server = server
```

```python
        self.key = key
        self.serialize = serialize_func

    @classmethod
    def from_settings(cls, settings):
        params = {
            'server': connection.from_settings(settings),
        }
        if settings.get('REDIS_ITEMS_KEY'):
            params['key'] = settings['REDIS_ITEMS_KEY']
        if settings.get('REDIS_ITEMS_SERIALIZER'):
            params['serialize_func'] = load_object(
                settings['REDIS_ITEMS_SERIALIZER']
            )

        return cls(**params)

    @classmethod
    def from_crawler(cls, crawler):
        return cls.from_settings(crawler.settings)

    def process_item(self, item, spider):
        return deferToThread(self._process_item, item, spider)

    def _process_item(self, item, spider):
        key = self.item_key(item, spider)
        data = self.serialize(item)
        self.server.rpush(key, data)
        return item

    def item_key(self, item, spider):
        """Returns redis key based on given spider.

        Override this function to use a different key depending on the item
        and/or spider.

        """
        return self.key % {'spider': spider.name}
```

- self.server 是 Redis 数据库的连接对象，与 FifoQueue 中相同。
- 观察在 self.server 上调用的方法可知，RedisPipeline 使用 Redis 中的一个列表存储所有爬虫爬取到的数据，该列表在数据库中的键为调用 item_key 方法的结果，即 self.key % {'spider': spider.name}。self.key 可以通过配置文件设置（REDIS_ITEMS_KEY），默认情况下列表的键为<spider_name>:items。
- Redis 的列表只能存储字符串，而 Spider 爬取到的数据 item 的类型是 Item 或 Python 字典，所以先要将 item 串行化成字符串，再存入 Redis 列表。串行化函数也可以通过配置文件指定（REDIS_ITEMS_SERIALIZER），默认情况下 item 被串行化成 json 串。
- process_item 方法处理爬取到的每一项数据，因为写入数据库为 I/O 操作，速度较慢，所以可以在线程中执行，调用 twisted 中的 deferToThread 方法，启动线程执行 _process_item 方法。
- _process_item 方法实际处理爬取到的每一项数据，先使用 self.serial 函数将 item 串行化成字符串，再调用 Redis 的 rpush 命令将其写入数据库中的列表。

14.3 使用 scrapy-redis 进行分布式爬取

了解了 scrapy-redis 的原理后，我们学习使用 scrapy + scrapy-redis 进行分布式爬取。

14.3.1 搭建环境

首先搭建 scrapy-redis 分布式爬虫环境，当前我们有 3 台 Linux 主机。

云服务器（A）：116.29.35.201 (Redis Server)
云服务器（B）：123.59.45.155
本机（C）：1.13.41.127

在 3 台主机上安装 scrapy 和 scrapy-redis：

```
$ pip install scrapy
$ pip install scrapy-redis
```

选择其中一台云服务器搭建供所有爬虫使用的 Redis 数据库，步骤如下：

步骤 01 在云服务器上安装 redis-server。
步骤 02 在 Redis 配置文件中修改服务器的绑定地址（确保数据库可被所有爬虫访问）。
步骤 03 启动/重启 Redis 服务器。

登录云服务器（A），在 bash 中完成上述步骤：

```
116.29.35.201$ sudo apt-get install redis-server
116.29.35.201$ sudo vi /etc/redis/redis.conf
...
# bind 127.0.0.1
bind 0.0.0.0
...
116.29.35.201$ sudo service redis-server restart
```

最后，在 3 台主机上测试能否访问云服务器（A）上的 Redis 数据库：

```
$ redis-cli -h 116.29.35.201 ping
PONG
```

到此，Scrapy 分布式爬虫环境搭建完毕。

14.3.2　项目实战

本章的核心知识点是分布式爬取，因此本项目不再对分析页面、编写 Spider 等大家熟知的技术进行展示。我们可以任意挑选一个在之前章节中做过的项目，将其改为分布式爬取的版本，这里以第 8 章的 toscrape_book 项目（爬取 books.toscrape.com 中的书籍信息）为例进行讲解。

复制 toscrape_book 项目，得到新项目 toscrape_book_distributed：

```
$ cp -r toscrape_book toscrape_book_distributed
$ cd toscrape_book_distributed
```

在配置文件 settings.py 中添加 scrapy-redis 的相关配置：

```
# 必选项
# ================================================================
# 指定爬虫所使用的 Redis 数据库（在云服务器 116.29.35.201 上）
REDIS_URL = 'redis://116.29.35.201:6379'

# 使用 scrapy_redis 的调度器替代 Scrapy 原版调度器
SCHEDULER = "scrapy_redis.scheduler.Scheduler"

# 使用 scrapy_redis 的 RFPDupeFilter 作为去重过滤器
DUPEFILTER_CLASS = "scrapy_redis.dupefilter.RFPDupeFilter"
```

```
# 启用 scrapy_redis 的 RedisPipeline 将爬取到的数据汇总到 Redis 数据库
ITEM_PIPELINES = {
    'scrapy_redis.pipelines.RedisPipeline': 300,
}
# 可选项
# ==============================================================
# 爬虫停止后，保留/清理 Redis 中的请求队列以及去重集合
# True：保留，False：清理，默认为 False
SCHEDULER_PERSIST = True
```

将单机版本的 Spider 改为分布式版本的 Spider，只需做如下简单改动：

```
from scrapy_redis.spiders import RedisSpider

# 1.更改基类
# class BooksSpider(spider.Spider):
class BooksSpider(RedisSpider):
    ...
    # 2.注释 start_urls
    #start_urls = ['http://books.toscrape.com/']
    ...
```

上述改动针对"如何为多个爬虫设置起始爬取点"这个问题，解释如下：

- 在分布式爬取时，所有主机上的代码是相同的，如果使用之前单机版本的 Spider 代码，那么每一台主机上的 Spider 都通过 start_urls 属性定义了起始爬取点，在构造起始爬取点的 Request 对象时，dont_filter 参数设置为了 True，即忽略去重过滤器的过滤。因此多个（数量等于爬虫数量）重复请求将强行进入 Redis 中的请求队列，这可能导致爬取到重复数据。
- 为了解决上述问题，scrapy-redis 提供了一个新的 Spider 基类 RedisSpider，RedisSpider 重写了 start_requests 方法，它尝试从 Redis 数据库的某个特定列表中获取起始爬取点，并构造 Request 对象（dont_filter=False），该列表的键可通过配置文件设置（REDIS_START_URLS_KEY），默认为<spider_name>:start_urls。在分布式爬取时，用户运行所有爬虫后，需手动使用 Redis 命令向该列表添加起始爬取点，之后只有其中一个爬虫能获取到起始爬取点，因此对应的请求也就只有一个，从而避免了重复。

到此，分布式版本的项目代码已经完成了，分发到各个主机：

```
$ scp -r toscrape_book_distributed liushuo@116.29.35.201:~/scrapy_book
$ scp -r toscrape_book_distributed liushuo@123.59.45.155:~/scrapy_book
```

分别在 3 台主机使用相同命令运行爬虫：

```
$ scrapy crawl books
2017-05-14 17:56:42 [scrapy.utils.log] INFO: Scrapy 1.3.3 started (bot: toscrape_book)
2017-05-14 17:56:42 [scrapy.utils.log] INFO: Overridden settings: {'DUPEFILTER_CLASS': 'scrapy_redis.dupefilter.RFPDupeFilter', 'FEED_EXPORT_FIELDS': ['upc', 'name', 'price', 'stock', 'review_rating', 'review_num'], 'SCHEDULER': 'scrapy_redis.scheduler.Scheduler', 'BOT_NAME': 'toscrape_book', 'ROBOTSTXT_OBEY': True, 'NEWSPIDER_MODULE': 'toscrape_book.spiders', 'SPIDER_MODULES': ['toscrape_book.spiders']}
2017-05-14 17:56:42 [scrapy.middleware] INFO: Enabled extensions:
['scrapy.extensions.logstats.LogStats',
 'scrapy.extensions.telnet.TelnetConsole',
 'scrapy.extensions.corestats.CoreStats']
2017-05-14 17:56:42 [books] INFO: Reading start URLs from redis key 'books:start_urls' (batch size: 16, encoding: utf-8
2017-05-14 17:56:42 [scrapy.middleware] INFO: Enabled downloader middlewares:
['scrapy.downloadermiddlewares.robotstxt.RobotsTxtMiddleware',
 'scrapy.downloadermiddlewares.httpauth.HttpAuthMiddleware',
 'scrapy.downloadermiddlewares.downloadtimeout.DownloadTimeoutMiddleware',
 'scrapy.downloadermiddlewares.defaultheaders.DefaultHeadersMiddleware',
 'scrapy.downloadermiddlewares.useragent.UserAgentMiddleware',
 'scrapy.downloadermiddlewares.retry.RetryMiddleware',
 'scrapy.downloadermiddlewares.redirect.MetaRefreshMiddleware',
 'scrapy.downloadermiddlewares.httpcompression.HttpCompressionMiddleware',
 'scrapy.downloadermiddlewares.redirect.RedirectMiddleware',
 'scrapy.downloadermiddlewares.cookies.CookiesMiddleware',
 'scrapy.downloadermiddlewares.stats.DownloaderStats']
2017-05-14 17:56:42 [scrapy.middleware] INFO: Enabled spider middlewares:
['scrapy.spidermiddlewares.httperror.HttpErrorMiddleware',
 'scrapy.spidermiddlewares.offsite.OffsiteMiddleware',
 'scrapy.spidermiddlewares.referer.RefererMiddleware',
 'scrapy.spidermiddlewares.urllength.UrlLengthMiddleware',
 'scrapy.spidermiddlewares.depth.DepthMiddleware']
2017-05-14 17:56:42 [scrapy.middleware] INFO: Enabled item pipelines:
```

```
['scrapy_redis.pipelines.RedisPipeline']
2017-05-14 17:56:42 [scrapy.core.engine] INFO: Spider opened
2017-05-14 17:56:42 [scrapy.extensions.logstats] INFO: Crawled 0 pages (at 0 pages/min), scraped 0 items (at 0 items/min)
2017-05-14 17:56:42 [scrapy.extensions.telnet] DEBUG: Telnet console listening on 127.0.0.1:6023
...阻塞在此处...
```

运行后,由于 Redis 中的起始爬取点列表和请求队列都是空的,3 个爬虫都进入了暂停等待的状态,因此在任意主机上使用 Redis 客户端设置起始爬取点:

```
$ redis-cli -h 116.29.35.201
116.29.35.201:6379> lpush books:start_urls 'http://books.toscrape.com/'
(integer) 1
```

随后,其中一个爬虫(本实验中是云服务器 A)从起始爬取点列表中获取到了 url,在其 log 中观察到如下信息:

```
2017-05-14 17:57:18 [books] DEBUG: Read 1 requests from 'books:start_urls'
```

该爬虫用起始爬取点 url 构造的 Request 对象最终被添加到 Redis 中的请求队列之后。各个爬虫相继开始工作了,可在各爬虫的 log 中观察到类似于如下的信息:

```
2017-05-14 18:00:42 [scrapy.core.scraper] DEBUG: Scraped from <200 http://books.toscrape.com/catalogue/arena_587/index.html>
    {'name': 'Arena',
     'price': '£21.36',
     'review_num': '0',
     'review_rating': 'Four',
     'stock': '11',
     'upc': '2c34f9432069b52b'}
2017-05-14 18:00:42 [scrapy.core.engine] DEBUG: Crawled (200)　(referer: http://books.toscrape.com/catalogue/page-21.html)
2017-05-14 18:00:42 [scrapy.core.scraper] DEBUG: Scraped from <200 http://books.toscrape.com/catalogue/adultery_586/index.html>
    {'name': 'Adultery',
     'price': '£20.88',
     'review_num': '0',
     'review_rating': 'Five',
     'stock': '11',
     'upc': 'bb967277222e689c'}
2017-05-14 18:00:42 [scrapy.core.engine] DEBUG: Crawled (200)　(referer:
```

http://books.toscrape.com/catalogue/page-21.html)

2017-05-14 18:00:42 [scrapy.core.scraper] DEBUG: Scraped from <200 http://books.toscrape.com/catalogue/a-mothers-reckoning-living-in-the-aftermath-of-tragedy_585/index.html>

{'name': "A Mother's Reckoning: Living in the Aftermath of Tragedy",
 'price': '£19.53',
 'review_num': '0',
 'review_rating': 'Three',
 'stock': '11',
 'upc': '2b69dec0193511d9'}

2017-05-14 18:00:43 [scrapy.core.scraper] DEBUG: Scraped from <200 http://books.toscrape.com/catalogue/112263_583/index.html>

{'name': '11/22/63',
 'price': '£48.48',
 'review_num': '0',
 'review_rating': 'Three',
 'stock': '11',
 'upc': 'a9d7b75461084a26'}

2017-05-14 18:00:43 [scrapy.core.engine] DEBUG: Crawled (200) (referer: http://books.toscrape.com/catalogue/page-21.html)

2017-05-14 18:00:43 [scrapy.core.scraper] DEBUG: Scraped from <200 http://books.toscrape.com/catalogue/10-happier-how-i-tamed-the-voice-in-my-head-reduced-stress-without-losing-my-edge-and-found-self-help-that-actually-works_582/index.html>

{'name': '10% Happier: How I Tamed the Voice in My Head, Reduced Stress '
         'Without Losing My Edge, and Found Self-Help That Actually Works',
 'price': '£24.57',
 'review_num': '0',
 'review_rating': 'Two',
 'stock': '10',
 'upc': '34669b2e9d407d3a'}

等待全部爬取完成后，在 Redis 中查看爬取到的数据：

```
116.29.35.201:6379> keys *
1) "books:items"
2) "books:dupefilter"
116.29.35.201:6379> llen books:items
(integer) 1000
116.29.35.201:6379> LRANGE books:items 0 4
```

1) "{\"stock\": \"22\", \"review_num\": \"0\", \"upc\": \"a897fe39b1053632\", \"name\": \"A Light in the Attic\", \"review_rating\": \"Three\", \"price\": \"\\u00a351.77\"}"

2) "{\"stock\": \"20\", \"review_num\": \"0\", \"upc\": \"e00eb4fd7b871a48\", \"name\": \"Sharp Objects\", \"review_rating\": \"Four\", \"price\": \"\\u00a347.82\"}"

3) "{\"stock\": \"20\", \"review_num\": \"0\", \"upc\": \"90fa61229261140a\", \"name\": \"Tipping the Velvet\", \"review_rating\": \"One\", \"price\": \"\\u00a353.74\"}"

4) "{\"stock\": \"20\", \"review_num\": \"0\", \"upc\": \"6957f44c3847a760\", \"name\": \"Soumission\", \"review_rating\": \"One\", \"price\": \"\\u00a350.10\"}"

5) "{\"stock\": \"19\", \"review_num\": \"0\", \"upc\": \"2597b5a345f45e1b\", \"name\": \"The Dirty Little Secrets of Getting Your Dream Job\", \"review_rating\": \"Four\", \"price\": \"\\u00a333.34\"}"

116.29.35.201:6379> LRANGE books:items -5 -1

1) "{\"name\": \"Shameless\", \"price\": \"\\u00a358.35\", \"review_rating\": \"Three\", \"upc\": \"c068c013d6921fea\", \"review_num\": \"0\", \"stock\": \"1\"}"

2) "{\"stock\": \"1\", \"review_num\": \"0\", \"upc\": \"19fec36a1dfb4c16\", \"name\": \"A Spy's Devotion (The Regency Spies of London #1)\", \"review_rating\": \"Five\", \"price\": \"\\u00a316.97\"}"

3) "{\"stock\": \"1\", \"review_num\": \"0\", \"upc\": \"f684a82adc49f011\", \"name\": \"1st to Die (Women's Murder Club #1)\", \"review_rating\": \"One\", \"price\": \"\\u00a353.98\"}"

4) "{\"stock\": \"1\", \"review_num\": \"0\", \"upc\": \"228ba5e7577e1d49\", \"name\": \"1,000 Places to See Before You Die\", \"review_rating\": \"Five\", \"price\": \"\\u00a326.08\"}"

5) "{\"name\": \"Girl in the Blue Coat\", \"price\": \"\\u00a346.83\", \"review_rating\": \"Two\", \"upc\": \"41fc5dce044f16f5\", \"review_num\": \"0\", \"stock\": \"3\"}"

如上所示，我们成功地爬取到了 1000 项数据（由各爬虫最后的 log 信息得知，爬虫 A:514 项，爬虫 B:123 项，爬虫 C:363 项）。每一项数据以 json 形式存储在 Redis 的列表中，需要使用这些数据时，可以编写 Python 程序将它们从 Redis 中读出，代码框架如下：

```python
import redis
import json

ITEM_KEY = 'books:items'

def process_item(item):
    # 添加处理数据的代码
    ...

def main():
    r = redis.StrictRedis(host='116.29.35.201', port=6379)
    for _ in range(r.llen(ITEM_KEY)):
```

```
            data = r.lpop(ITEM_KEY)
            item = json.loads(data.decode('utf8'))
            process_item(item)

if __name__ == '__main__':
    main()
```

到此，我们完成了分布式爬取的项目。

14.4 本章小结

本章我们学习了如何利用 scrapy-redis 构建分布式 Scrapy 爬虫，首先介绍了一些 Redis 数据库的基础知识，然后对 scrapy-redis 源码进行了分析，最后通过案例展示了一个分布式爬虫的开发流程。